AF352831

Broken Symmetry in Curved Spacetime and Gravity

Broken Symmetry in Curved Spacetime and Gravity

Special Issue Editor

Charles D. Lane

MDPI • Basel • Beijing • Wuhan • Barcelona • Belgrade • Manchester • Tokyo • Cluj • Tianjin

Special Issue Editor
Charles D. Lane
Berry College and Indiana University CSS
USA

Editorial Office
MDPI
St. Alban-Anlage 66
4052 Basel, Switzerland

This is a reprint of articles from the Special Issue published online in the open access journal *Symmetry* (ISSN 2073-8994) (available at: https://www.mdpi.com/journal/symmetry/special_issues/Broken_ Symmetry_Curved_Spacetime_Gravity).

For citation purposes, cite each article independently as indicated on the article page online and as indicated below:

LastName, A.A.; LastName, B.B.; LastName, C.C. Article Title. *Journal Name* **Year**, *Article Number*, Page Range.

ISBN 978-3-03936-449-7 (Hbk)
ISBN 978-3-03936-450-3 (PDF)

Contents

About the Special Issue Editor

Charles D. Lane (Associate Professor of Physics at Berry College and Visiting Professor of Physics (courtesy appointment) at Indiana University Center for Spacetime Symmetries) has studied Lorentz Violation for 25 years. He earned his doctorate in Mathematical Physics from Indiana University in 2000. Following that, he served a one-year appointment as a Faculty Fellow in Physics at Colby College in Maine, and then he began his long-term position at Berry College in Georgia in August 2001. He has also held a courtesy appointment as Visiting Professor at Indiana University as part of the Center for Spacetime Symmetries since January 2016.

Preface to "Broken Symmetry in Curved Spacetime and Gravity"

Modern physics rests on a foundation of two fundamental theories: General Relativity (GR) and the Standard Model (SM). Each theory agrees extremely well with experiments in a certain domain. However, the predictions of the theories disagree with each other in certain situations. Therefore, GR and the SM are likely to be low-energy approximations to some more fundamental theory. A major current goal in physics is to determine the nature of this more fundamental theory.

The most natural approach to learning about the fundamental theory is to look for situations where GR and the SM strongly disagree; in such situations, at least one of these theories must make predictions that are clearly wrong. Unfortunately, all known situations where the theories strongly disagree are untenable to study experimentally. This leads us to consider an alternate approach—suppose that one or both of these theories is slightly wrong in a situation where we can perform experiments with extremely high precision. Careful study of these high-precision experiments could reveal a violation of one of the current theories. This is the approach taken in studies of spacetime symmetry violation.

Charles D. Lane
Special Issue Editor

Article

Constraints on Lorentz Invariance Violation from Optical Polarimetry of Astrophysical Objects

Fabian Kislat

Department of Physics and Space Science Center, University of New Hampshire, 8 College Road, Durham, NH 03824, USA; fabian.kislat@unh.edu

Received: 7 September 2018; Accepted: 2 November 2018; Published: 5 November 2018

Abstract: Theories of quantum gravity suggest that Lorentz invariance, the fundamental symmetry of the Theory of Relativity, may be broken at the Planck energy scale. While any deviation from conventional Physics must be minuscule in particular at attainable energies, this hypothesis motivates ever more sensitive tests of Lorentz symmetry. In the photon sector, astrophysical observations, in particular polarization measurements, are a very powerful tool because tiny deviations from Lorentz invariance will accumulate as photons propagate over cosmological distances. The Standard-Model Extension (SME) provides a theoretical framework in the form of an effective field theory that describes low-energy effects due to a more fundamental quantum gravity theory by adding additional terms to the Standard Model Lagrangian. These terms can be ordered by the mass dimension d of the corresponding operator and lead to a wavelength, polarization, and direction dependent phase velocity of light. Lorentz invariance violation leads to an energy-dependent change of the Stokes vector as photons propagate, which manifests itself as a rotation of the polarization angle in measurements of linear polarization. In this paper, we analyze optical polarization measurements from 63 Active Galactic Nuclei (AGN) and Gamma-ray Bursts (GRBs) to search for Lorentz violating signals. We use both spectropolarimetric measurements, which directly constrain the change of linear polarization angle, as well as broadband spectrally integrated measurements. In the latter, Lorentz invariance violation manifests itself by reducing the observed net polarization fraction. Any observation of non-vanishing linear polarization thus leads to constraints on the magnitude of Lorentz violating effects. We derive the first set limits on each of the 10 individual birefringent coefficients of the minimal SME with $d = 4$, with 95 % confidence limits on the order of 10^{-34} on the dimensionless coefficients.

Keywords: Lorentz invariance; Standard-Model extension; polarization; Active Galactic Nuclei; Gamma-ray Bursts

---

1. Introduction

Lorentz invariance is the fundamental symmetry of Einstein's theory of Special Relativity. It has been established by many classic experiments, such as Michelson–Morley, Kennedy–Thorndike, and Ives–Stilwell [1–3], and tested to great precision by modern experiments [4]. Theories of quantum gravity suggest that there may be minute deviations from Lorentz symmetry, which motivates ever more sensitive tests [5–10].

Violations of Lorentz symmetry can lead to an energy-dependent vacuum photon dispersion relation, birefringence as well as anisotropy of the vacuum [11]. All of these effects can be tested with astrophysical observations, which are particularly sensitive because minuscule effects accumulate as photons propagate over very large distances resulting in measurable effects [12]. Vacuum birefringence leads to a wavelength-dependent change of the Stokes parameters, generally resulting in a rotation of the linear polarization angle. Astrophysical tests of Lorentz symmetry include time of flight measurements (see, e.g., Refs. [13–17]) and polarization measurements (see, e.g., Refs. [18–20]). The latter are generally more sensitive than time-of-flight measurements by the ratio between the period of the light wave and the time resolution of dispersion tests, which is usually limited not by the time resolution of the detector but by source-dependent flux variability time scales and photon statistics.

The Standard-Model Extension (SME) is an effective field theory framework describing low-energy effects of a more fundamental theory of quantum gravity, including violations of Lorentz and CPT invariance [11,12]. It introduces additional terms in the Standard-Model Lagrangian, which can in part be ordered by the mass dimension d of the underlying operator. Terms with $d \geq 5$ are non-renormalizable and generally thought to be suppressed by $M_{\mathrm{Planck}}^{4-d}$, whereas the minimal SME with renormalizable operators of $d \leq 4$ results in effects that are unsuppressed relative to conventional physics, unless some hierarchy of scales exists. Operators of even d are CPT even, while odd-d operators are CPT odd. In the photon sector, there are $(d-1)^2$ non-birefringent coefficients and $2(d-1)^2 - 8$ birefringent coefficients describing photon propagation in vacuum for even d. For odd d, there are $(d-1)^2$ birefringent coefficients. In general, these coefficients result in an anisotropy of the vacuum, and time of flight or polarization measurements of a single astrophysical source can only constrain combinations of these coefficients. A notable exception are measurements of the polarization of the Cosmic Microwave Background, which resulted in extremely tight constraints on all coefficients of $d = 3$ [21–23]. In previous papers, we used gamma-ray time-of-flight measurements and optical polarization measurements from multiple sources to individually constrain all non-birefringent parameters of mass-dimension $d = 6$ [16] and all parameters with $d = 5$ [20], respectively.

Here, we use the same optical polarization measurements of Active Galactic Nuclei (AGN) and Gamma-ray Bursts (GRBs) as in Ref. [20] to individually constrain each of the 10 birefringent coefficients with mass-dimension $d = 4$. The non-birefringent coefficients of $d = 4$ result in an energy-independent anisotropy of the phase velocity of light. Laboratory searches using a variety of resonating cavities have resulted in strong constraints on each of these coefficients [4], with the strongest constraints from long-baseline gravitational wave interferometers [24]. On the other hand, there are very few constraints of combinations of the birefringent coefficients, mostly from X-ray polarimetry of GRBs [19].

The main challenge, compared to the analysis of the $d = 5$ case, is that unlike odd d at even d the change of the linear polarization during propagation depends on the linear polarization angle. Therefore, the analysis requires a different approach. As before, we make use both of polarization measurements integrated over the relatively broad bandwidth of a telescope, as well as spectropolarimetric measurements, which provide the Stokes parameters Q and U as a function of wavelength.

The paper is structured as follows. In Section 2, we summarize the theoretical background and derive expressions for the observable effects that will serve as the foundation for the data analysis. In Section 3, we lay out the analysis of both spectropolarimetric and spectrally integrated polarization measurements and their interpretation in terms of the SME. In Section 4, we describe the Markov-Chain Monte Carlo method we use to derive limits on the individual SME coefficients and give the results. Finally, in Section 5, we discuss these results and give an outlook to future possibilities. Furthermore, Appendices A and B list the astrophysical sources used in this analysis and distributions of the different SME coefficients derived from these measurements, respectively.

2. Theory

The photon vacuum dispersion relation of the Standard Model Extension (SME) can be written as [12]

$$E \simeq \left(1 - \varsigma^0 \pm \sqrt{(\varsigma^1)^2 + (\varsigma^2)^2 + (\varsigma^3)^2} \right) p \tag{1}$$

with the expansion in spin-weighted spherical harmonics $_sY_{jm}$ and mass dimension d:

$$\varsigma^0 = \sum_{djm} E^{d-4}\, _0Y_{jm}(\hat{n}) c^{(d)}_{(I)jm},$$

(2)

$$\varsigma^{\pm} = \varsigma^1 \mp i\varsigma^2 = \sum_{djm} E^{d-4}\, _{\pm 2}Y_{jm}(\hat{n}) \left(k^{(d)}_{(E)jm} \pm ik^{(d)}_{(B)jm}\right),$$

(3)

$$\varsigma^3 = \sum_{djm} E^{d-4}\, _0Y_{jm}(\hat{n}) k^{(d)}_{(V)jm},$$

(4)

where $\hat{n}$ is the direction towards the origin of the photon. For odd d, there are $(d-1)^2$ coefficients $k^{(d)}_{(V)jm}$ and, for even d, there are $(d-1)^2$ non-birefringent coefficients $c^{(d)}_{(I)jm}$ and $(d-1)^2 - 4$ birefringent coefficients $k^{(d)}_{(E)jm}$ and $k^{(d)}_{(B)jm}$ each. In this paper, we restrict our attention to the birefringent coefficients with $d = 4$ in the minimal SME, i.e.,

$$\varsigma^{\pm}\big|_{d=4} = \sum_{m=-2}^{+2} {}_{\pm 2}Y_{2m}(\hat{n}) \left(k^{(4)}_{(E)2m} \pm ik^{(4)}_{(B)2m}\right)$$

(5)

with a total of 10 coefficients $k^{(4)}_{(E)2m}$ and $k^{(4)}_{(B)2m}$, which comprise a total of 10 real components since

$$k^{(d)}_{(E,B)j(-m)} = (-1)^m \left(k^{(d)}_{(E,B)jm}\right)^*.$$

(6)

The polarization of an electromagnetic wave is completely described by the four Stokes parameters: intensity I; linear polarization Q and U, where U describes linear polarization at an angle of $45°$ relative to Q; and circular polarization V. General elliptical polarization is described by the Stokes vector $s = (Q, U, V)^T$. The polarization of photons with energy E will change as they propagate through a birefringent vacuum:

$$\frac{ds}{dt} = 2E\varsigma \times s$$

(7)

with the birefringence axis $\varsigma = (\varsigma^1, \varsigma^2, \varsigma^3)^T$. In the CPT-odd case, this axis is aligned with the V axis and as a result, linearly polarized light remains linearly polarized, but the polarization position angle will rotate as light propagates.

In the CPT-even case, the birefringence axis lies in the $Q - U$ plane. Consequently, the Stokes vector will generally rotate out of this plane and linearly polarized light will become elliptically polarized during propagation. However, light with s parallel to ς will remain unaffected. The eigenmode of propagation is described by the polarization angle (following the convention used in Ref. [12])

$$\xi/2 = \frac{1}{2} \arctan\left(\frac{-\varsigma^2}{\varsigma^1}\right).$$

(8)

The observed polarization of light emitted by a source at redshift z can conveniently be calculated in a spin-weighted Stokes basis $s = (s_{(+2)}, s_{(0)}, s_{(-2)})^T$ with $s_{(0)} = V$ and $s_{(\pm 2)} = Q \mp iU$ and $\varsigma = (\varsigma^+, \varsigma^3, \varsigma^-)^T$. Then, the observed Stokes vector s is related to the blueshifted Stokes vector s_z by [12]

$$s = \mathbf{M}_z \cdot s_z$$

(9)

with the Müller matrix

$$\mathbf{M}_z = \begin{pmatrix} \cos^2\Phi_z & -i\sin(2\Phi_z)e^{-i\xi} & \sin^2\Phi_z e^{-2i\xi} \\ -\frac{i}{2}\sin(2\Phi_z)e^{i\xi} & \cos(2\Phi_z) & \frac{i}{2}\sin(2\Phi_z)e^{-i\xi} \\ \sin^2\Phi_z e^{2i\xi} & i\sin(2\Phi_z)e^{i\xi} & \cos^2\Phi_z \end{pmatrix}$$

(10)

and, at $d = 4$,

$$\Phi_z = E \int_0^z \frac{dz'}{H_{z'}} \left| \sum_m {}_2Y_{2m}(\hat{\boldsymbol{n}}) \left(k^{(4)}_{(E)2m} \pm i k^{(4)}_{(B)2m} \right) \right|. \tag{11}$$

For convenience, we define the following abbreviations:

$$S(\hat{\boldsymbol{n}}) = \sum_m {}_2Y_{2m}(\hat{\boldsymbol{n}}) \left(k^{(4)}_{(E)2m} \pm i k^{(4)}_{(B)2m} \right), \tag{12}$$

$$L_z = \int_0^z \frac{dz'}{H_{z'}}, \tag{13}$$

$$\gamma(\hat{\boldsymbol{n}}) = |S(\hat{\boldsymbol{n}})|, \tag{14}$$

$$\vartheta_z(\hat{\boldsymbol{n}}) = L_z \gamma(\hat{\boldsymbol{n}}), \tag{15}$$

so that

$$\Phi_z = E \vartheta_z(\hat{\boldsymbol{n}}) \tag{16}$$

and

$$\xi = \arctan \left(\frac{-\Im(S)}{\Re(S)} \right). \tag{17}$$

Most astrophysically relevant emission mechanisms are not expected to produce any significant circular polarization. Hence, assuming 100 % linearly polarized light at the source with the polarization angle ψ_z,

$$Q_z = \cos(2\psi_z) \qquad U_z = \sin(2\psi_z) \qquad V_z = 0, \tag{18}$$

the observer Stokes parameters are

$$Q = \cos(2\psi_z) \cos^2\left(E\vartheta_z(\hat{\boldsymbol{n}}) \right) + \cos(2(\xi - \psi_z)) \sin^2\left(E\vartheta_z(\hat{\boldsymbol{n}}) \right) \tag{19}$$

$$U = \sin(2(\xi - \psi_z)) \sin^2\left(E\vartheta_z(\hat{\boldsymbol{n}}) \right) + \sin(2\psi_z) \cos^2\left(E\vartheta_z(\hat{\boldsymbol{n}}) \right). \tag{20}$$

Since the data used in this analysis do not contain any information about circular polarization, we do not consider V. The change in Stokes parameters is:

$$\Delta Q = Q - Q_z = -2\sin^2\left(E\vartheta_z(\hat{\boldsymbol{n}}) \right) \sin\xi \sin(\xi - 2\psi_z), \tag{21}$$

$$\Delta U = U - U_z = 2\sin^2\left(E\vartheta_z(\hat{\boldsymbol{n}}) \right) \cos\xi \sin(\xi - 2\psi_z). \tag{22}$$

The above expressions can be further simplified by realizing that the reference direction for the polarization angle can be chosen freely. Rotating the coordinate system such that $\xi' = 0$, we express all position angles as $\psi' = \psi - \xi/2$ and find

$$\Delta Q' = 0, \tag{23}$$

$$\Delta U' = -2\sin^2\left(E\vartheta_z(\hat{\boldsymbol{n}}) \right) U_z'. \tag{24}$$

All primed quantities are expressed in this rotated frame, while all polarization angles without prime are given in a frame where a polarization angle of 0 corresponds to linear polarization in the north/south direction and 90° to the east/west direction [25]. Quantities with subscript z refer to the polarization of the source at redshift z, quantities without a subscript to the observer polarization predicted by the SME, and quantities with subscript m, in the following, refer to measured quantities.

3. Astrophysical Polarization Measurements

Essentially, birefringence leads to an energy and ψ_z-dependent rotation of the polarization angle and change in linear polarization fraction, as illustrated in Figures 1 and 2. By measuring the linear polarization of photons emitted by distant objects, strong constraints on birefringence can be obtained.

In the analysis presented here, we make use of two kinds of measurements: spectropolarimetric measurements, where polarization fraction and angle are measured as a function of photon energy, and spectrally integrated measurements, where the polarization fraction is measured by integrating over a broad bandwidth determined by a filter in the optical path. Both analyses are based on the results from the previous section, but proceed differently. The goal of this section is to develop statistical measures for each type of observation that allows us to quantify the compatibility of a set of SME coefficients with the observation. The results are then combined in Section 4 into a joint probability function that is used to derive confidence intervals for each individual SME coefficient.

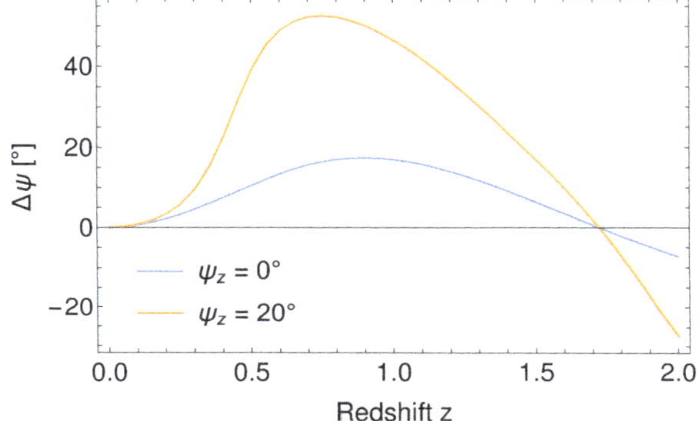

Figure 1. Difference of the observed polarization angle $\Delta\psi$ between photons of observer wavelengths of 1033 nm and 443 nm that were emitted with the same linear polarization angle $\psi_z = 0$ (blue) and $\psi_z = 20°$ (orange) as a function of source red shift. The photons arrive from the direction of GRB 990510, and $\Re(k^{(4)}_{(E)21}) = 10^{-32}$ with all other SME coefficients set to 0, resulting in an angle of the eigenmode in the $Q - U$ plane of the Stokes space of $\xi = -24.8°$.

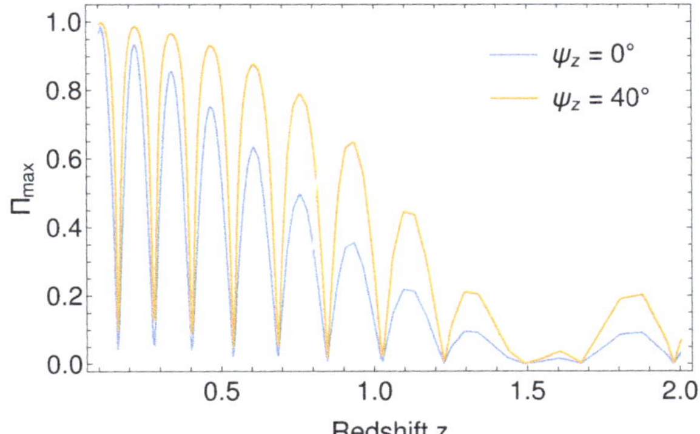

Figure 2. Effective polarization of light observed through the HOWPol R-band filter arriving from the direction of GRB 091208B if the light at the source is 100 % linearly polarized with a wavelength-independent polarization angle as a function of source redshift. For this illustration, the SME coefficients were set to 0, except for $\Re(k^{(4)}_{(E)21}) = 2 \times 10^{-32}$. The depolarization is due to averaging over the bandwidth of the filter and the rotation of the polarization angle, as shown for example in Figure 1, as well as the change of linear into circular polarization described by the Müller matrix (Equation (10)). The combination of these two effects leads to the observed "ringing". As described in Section 2, the effect depends on the linear polarization angle ψ_z at the source.

3.1. Spectropolarimetry

When measuring the polarization angle as a function of energy, $\psi(E)$, we can directly compare the result to the position angle resulting from Equations (19) and (20). Here, we reduce the problem to comparing the change in polarization angle at a given wavelength as predicted by the SME given a set of coefficients to the observed change over an instrument band pass. We start with a linear fit of the measured polarization angle as function of energy,

$$\psi_m(E) = \psi_m(\bar{E}) + \rho_m(E - \bar{E}), \tag{25}$$

where $\psi_m(\bar{E})$ is the measured polarization angle at the median energy $\bar{E} = 2.26\,\text{eV}$ of the fit range, and ρ_m with the uncertainty σ_ρ is the linear rate of change of polarization angle as a function of energy. An example fit is shown in Figure 3.

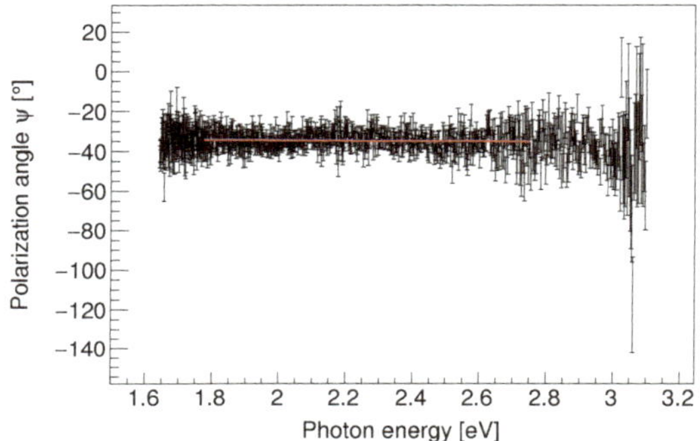

Figure 3. Example of a spectropolarimetric measurement. The figure shows the polarization angle of 4C 14.23 as a function of photon energy observed on 23 November 2009. The red line is a linear fit in order to determine ρ_m for comparison with Equation (28) to calculate the probability in Equation (29). All fit results are listed in Table A3.

To compare the measured rate of change $\rho_m \pm \sigma_\rho$ with the prediction due to a given set of SME parameters, we first find the source polarization angle ψ_z' from the observed polarization angle $\psi_m(\bar{E})$ at the median energy of the detector band pass. From Equations (23) and (24), we have

$$Q_z' = Q_m'(\bar{E}) \quad \text{and} \quad U_z' = \frac{U_m'(\bar{E})}{\cos(2\bar{E}\vartheta)}, \tag{26}$$

where $Q_m'(\bar{E}) = \cos(2(\psi_m(\bar{E}) - \xi/2))$ and $U_m'(\bar{E}) = \sin(2(\psi_m(\bar{E}) - \xi/2))$, so that

$$\psi_z' = \frac{1}{2}\arctan\left(\frac{U_z'}{Q_z'}\right). \tag{27}$$

Then, we linearize the predicted change in polarization angle with energy as given by the Stokes parameters in Equations (19) and (20), which is adequate for small changes over the bandwidth:

$$\bar{\rho} := \frac{d\psi}{dE}(\bar{E}) = \frac{d\psi'}{dE}(\bar{E}) = \frac{\vartheta\sin(2\bar{E}\vartheta)\sin(4\psi_z')}{2\left(\cos^2(2\psi_z') + \cos^2(\bar{E}\vartheta)\sin^2(2\psi_z')\right)}. \tag{28}$$

An example of the rotation of the linear polarization angle as a function of one of the SME coefficients is shown in Figure 4. The observer polarization angle in general oscillates around the source polarization angle and the roots of $\bar{\rho}$ seen in the figure correspond to values of $k^{(4)}_{(E)20}$ for which this oscillation is extremal at $\bar{E}$.

This allows us to quantify the compatibility of the measurement with a given set of SME coefficients. The probability to observe a change of polarization angle $\rho < |\rho_m|$ assuming a true change $|\bar{\rho}|$ given by Lorentz violation according to Equation (28) and given the uncertainty of the measurement, σ_ρ, is:

$$P(\rho < |\rho_m| \,|\, |\bar{\rho}|, \sigma_\rho) = \int_{-\infty}^{|\rho_m|} \frac{1}{\sqrt{2\pi\sigma_m^2}} \exp\left(-\frac{(\rho - |\bar{\rho}|)^2}{2\sigma_m^2} \right) d\rho. \tag{29}$$

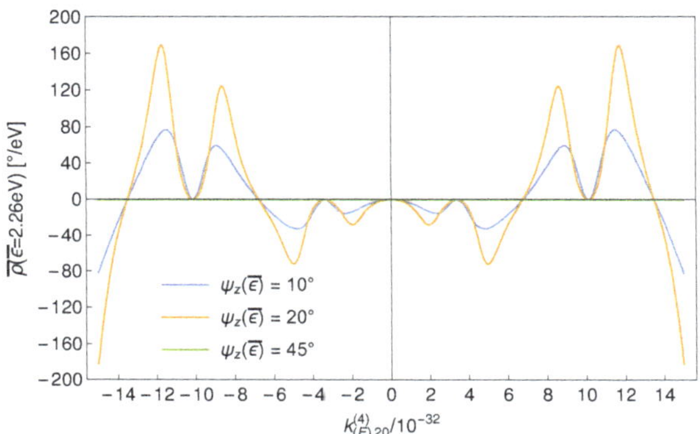

Figure 4. Rotation of the linear polarization angle with energy according to Equation (28) as a function of the SME coefficient $k^{(4)}_{(E)20}$ while keeping all other SME coefficients at 0. The source was assumed to be at redshift $z = 1$ with a codeclination of $\theta = 90°$. Results are shown for three different values of the polarization angle at the source, ψ_z. For $\psi_z = 45°$, the Stokes vector rotates out of the $Q - U$ plane, but the linear polarization angle does not change because $Q'_z = 0$ (see Equations (26) and (27)).

An example of this probability as a function of one of the SME coefficients is shown in Figure 5. The spikes are due to the roots of $\bar{\rho}$, but their width decreases with increasing SME coefficients. A single observation cannot be used to place constraints on the SME coefficients, unless additional assumptions are made. However, combining multiple observations will lead to tight constraints roughly corresponding to the width of the central peak, because the spikes at larger values of the SME coefficients will not line up for different sources.

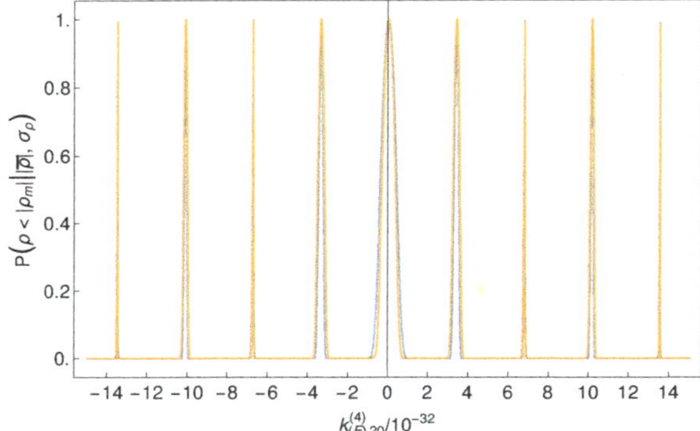

Figure 5. Probability $P(\rho < |\rho_m| \,||\, |\bar{\rho}|, \sigma_\rho)$ according to Equation (29) as a function of the SME coefficient $k^{(4)}_{(E)20}$ while keeping all other SME coefficients at 0. A measured change in polarization of $\rho_m = (0 \pm 1)\,°/\text{eV}$ was assumed. Source distance and direction are the same as in Figure 4, and colors have the same meaning. The roots of $\bar{\rho}$ in Figure 4 lead to the spikes in the probability function seen here.

The interpretation of the probability P is as follows. If $|\rho_m| < |\bar{\rho}|$, a strong degree of cancellation of the LIV-induced change of position angle with a source intrinsic change in position angle must have occurred, resulting in a low probability and a strong constraint on the given value of $\bar{\rho}$. On the other hand, if $|\rho_m| > |\bar{\rho}|$, there must be a strong source-intrinsic change of the polarization angle, irrespective of $\bar{\rho}$. Hence, P will be large and only a very weak constraint is placed on the given value of $\bar{\rho}$. This has deliberately been designed such that no claims of detection of Lorentz invariance violation will be made, because we do not want to make any assumptions about source-intrinsic energy-dependent changes of the polarization angle.

We applied this method to a large sample of publicly available spectropolarimetric measurements of AGN [26] covering observer frame wavelengths between 4000 Å and 7550 Å. From all data published in this archive until 30 March 2016, we selected all sources with redshift $z > 0.6$, which have at least one observation with a spectrally averaged polarization fraction $>10\,\%$. For each source, we chose the measurement that resulted in the largest average polarization fraction. We then fitted each polarization angle measurement as a function (25) of photon energy in the range 1.77 eV to 2.76 eV with a linear function centered at the median energy of this range in order to determine the change of polarization angle ρ_m. The resulting dataset was further reduced by removing all sources with $\rho_m/\sigma_\rho > 3$. The final list of measurements is given in Appendix A (Table A3).

3.2. Polarimetry Integrated over a Broad Bandwidth

When integrating over the bandpass of a broadband polarimeter, vacuum birefringence will lead to a reduction of the observed polarization compared to the polarization at the source due to the rotation of the polarization angle. We derive the largest possibly observable linear polarization fraction for an instrument with energy-dependent detection efficiency $T(E)$, assuming that the emitted light is 100 % linearly polarized with an energy-independent polarization angle ψ_z. We then quantify the compatibility of this maximum possible polarization with measured polarization fractions and angles. Any energy dependence of ψ_z will lead to an additional reduction of measured polarization, making this a conservative approach. While it is in principle possible that birefringence and source-intrinsic effects cancel, this is very unlikely, in particular when observing multiple astrophysical sources.

We start by computing the effective Stokes parameters of a measurement for a given set of SME coefficients and a 100 % polarized source. Additivity of Stokes parameters allows us to integrate them over the detector bandpass,

$$Q' = \int T(E)Q'(E)dE = \int T(E)(Q'_z + \Delta Q'(E))dE = Q'_z \int T(E)dE = \mathcal{N}Q'_z \tag{30}$$

and

$$\mathcal{U}' = \int T(E)U'(E)dE = \int T(E)(U'_z + \Delta U'(E))dE$$
$$= U'_z \left(\int T(E)dE - 2\int T(E)\sin^2(E\vartheta_z(\hat{n}))dE \right) = U'_z \mathcal{N}(1 - \mathcal{F}(\vartheta_z(\hat{n}))), \tag{31}$$

where we use the definitions in Equations (23) and (24) and introduce the instrument-dependent normalization constant

$$\mathcal{N} = \int T(E)dE \tag{32}$$

and the instrument-dependent function

$$\mathcal{F}(\vartheta) = \frac{2}{\mathcal{N}} \int T(E)\sin^2(E\vartheta)dE. \tag{33}$$

These integrals must be computed numerically, since $T(E)$ is typically measured for each individual instrument. The advantage of formulating the problem in this way, however, is that $\mathcal{N}$ solely depends on the instrument being used, and $\mathcal{F}(\vartheta)$ can be tabulated for efficient evaluation. The integrals in Equations (32) and (33) were calculated in the range 1.2 eV to 2.8 eV. All source properties (distance and direction) and SME coefficients are combined into the single instrument-independent parameter ϑ.

In this analysis, we used data from various optical telescopes employing a variety of filters. The filter transmission curves used in this analysis are shown in Figure 6. Table 1 lists the resulting normalization constants $\mathcal{N}$ and Figure 7 shows the tabulated functions $\mathcal{F}(\vartheta)$. With those definitions, the maximum observable polarization for a 100 % linearly polarized source is

$$\Pi_{\max} = \frac{\sqrt{Q'^2 + U'^2}}{\mathcal{N}} = \sqrt{Q'^2_z + U'^2_z (1 - \mathcal{F}(\vartheta))^2} = \sqrt{1 - U'^2_z \mathcal{F}(\vartheta)(2 - \mathcal{F}(\vartheta))}, \tag{34}$$

where we used $Q'^2_z + U'^2_z = 1$ in the last step. The corresponding observed polarization angle is

$$\Psi' = \Psi - \zeta/2 = \frac{1}{2}\arctan\left(\frac{\mathcal{U}'}{\mathcal{Q}'}\right) = \frac{1}{2}\arctan\left(\frac{U'_z(1 - \mathcal{F}(\vartheta))}{\pm\sqrt{1 - U'^2_z}}\right), \tag{35}$$

where the sign in the denominator is chosen to match the sign of Q'_z.

Table 1. Integral of the filter transmission curves used in this analysis (see Equation (32)).

Instrument	Filter	$\mathcal{N}\ [10^{-10}\,\text{GeV}]$
FORS1	R-band	3.840
FORS1	V-band	3.926
FORS2	R$_{\text{Special}}$	4.548
ALFOSC	R-band	3.135
EFOSCV	V-band	3.763
HOWPol	R-band	3.611

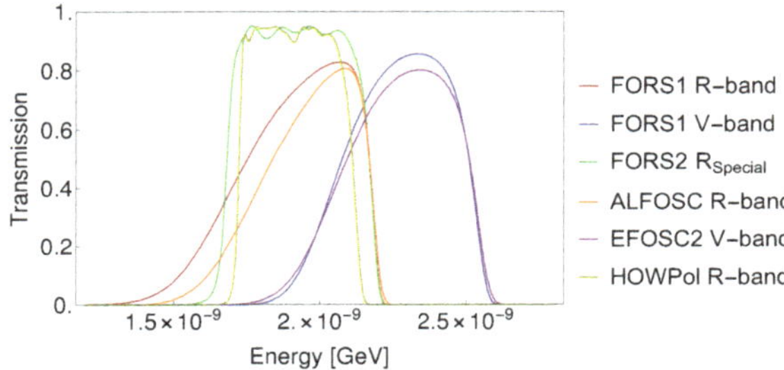

Figure 6. Filter transmission as a function of photon energy for the instruments and filters used in this analysis [27–31].

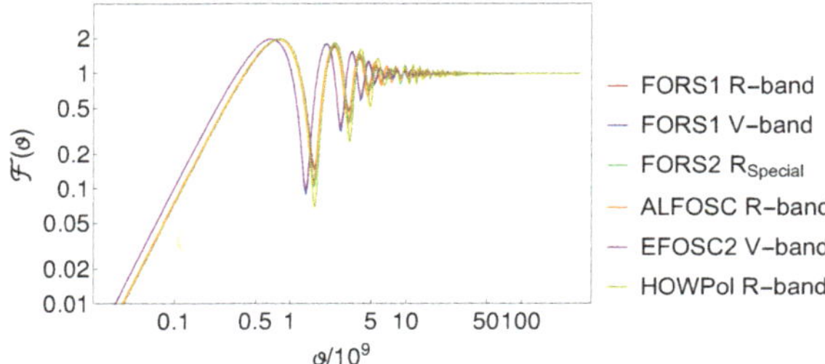

Figure 7. Instrument-dependent modulation integrals $\mathcal{F}(\vartheta)$ according to Equation (33) for the filter transmission curves shown in Figure 6.

The function $\mathcal{F}(\vartheta)$ determines the reduction of the observable linear polarization. It shows only a minor dependence on the exact shape of the transmission curve (compare for example the FORS1 R-band and the FORS2 R_{Special} filters), but a clear dependence on the energy range covered by the filter can be seen: for small values of ϑ, the V-band filters lead to larger values of $\mathcal{F}(\vartheta)$ than the R-band filters, and hence stronger sensitivity to Lorentz invariance violation. It is easy to show that in the limit of large SME coefficients

$$\lim_{\vartheta \to \infty} \mathcal{F}(\vartheta) = 1 \tag{36}$$

and it is obvious from Figure 7 that $\mathcal{F}(\vartheta)$ oscillates around this value as ϑ increases. In these cases, i.e., for $\mathcal{F}(\vartheta) = 1$, one finds

$$\Psi' = \begin{cases} \frac{\pi}{2} & \frac{\pi}{4} \le \psi'_z < \frac{3\pi}{4} \\ 0 & \text{else} \end{cases} \quad \text{and} \quad \Pi_{\max} = \sqrt{1 - U_z'^2}. \tag{37}$$

This result implies that only certain effective polarization angles Ψ can be observed in case ϑ is large. Hence, the observed polarization angle itself already places constraints on the SME coefficients.

Given a set of SME coefficients, source distance L_z and direction $\hat{n}$, we use Equations (34) and (35) to find the largest possible polarization fraction $\Pi_{\max}$ given a measured polarization angle Ψ_m. It is important to note that this does not imply the assumption of 100 % polarization at the source, but simply reflects the fact that we do not want to make any assumptions about the astrophysics of the source. In realistic models, the polarization of optical emission from blazars is expected to be at most on the order of 20–30 % [32,33]. Using such models as input, significantly tighter constraints would be possible. However, optical polarization of blazars is highly variable (see, e.g., [34]), and in case of GRB afterglows significantly higher degrees of polarization are possible [35]. To be conservative and to avoid systematic uncertainties, we decided against using detailed source models, and follow the approach used in previous work [19,20].

We numerically solve Equation (35) for U'_z by requiring $\Psi' = \Psi'_m$, where $\Psi'_m = \Psi_m - \zeta/2$ is the measured polarization angle in the rotated frame. Figure 8 illustrates this for different values of $\mathcal{F}(\vartheta)$ as function of the observed polarization angle Ψ'_m. The result is then used to calculate $\Pi_{\max}$ from Equation (34), shown as a function of Ψ'_m and $\mathcal{F}(\vartheta)$ in Figure 9. These results were also tabulated in the range $0.01 \leq \mathcal{F}(\vartheta) \leq 1$ in steps of $\delta\mathcal{F}(\vartheta) = 0.005$ and $\delta\Psi'_m = 1°$ for fast lookup at later stages of the analysis. Values of $\Pi_{\max}$ for arbitrary $\mathcal{F}(\vartheta)$ and Ψ'_m can be found from this table using bilinear interpolation. Figure 10 shows an example of the maximum theoretically possible net polarization for GRB 091208B as a function of one of the SME coefficients.

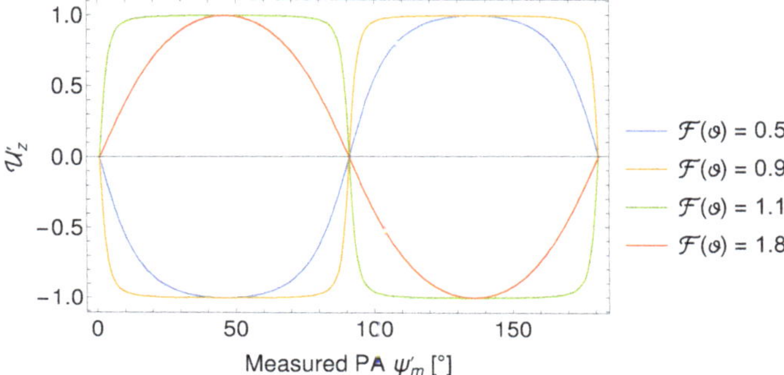

Figure 8. Source Stokes parameter U'_z as a function of Ψ_m calculated by numerically solving Equation (35) for $\Psi' = \Psi'_m$ for different values of the function $\mathcal{F}(\vartheta)$.

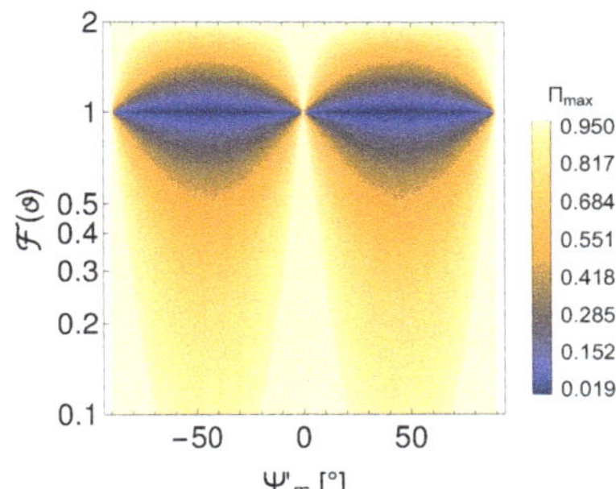

Figure 9. Maximum observable polarization as a function of measured polarization angle Ψ'_m and the oscillation integral $\mathcal{F}(\vartheta)$.

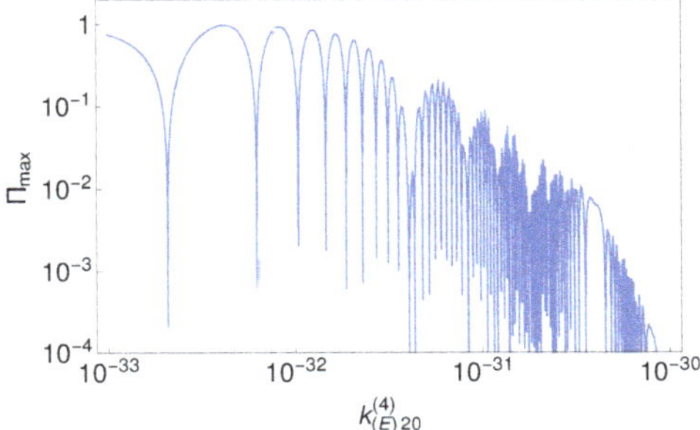

Figure 10. Maximum observable polarization fraction of GRB 091208B as a function of the SME coefficient $k^{(4)}_{(E)20}$ when keeping all other coefficient at 0.

The probability to observe a polarization Π given a true polarization $\hat{\Pi}$, can be written as [36,37]:

$$
\begin{aligned}
P(\Pi|\hat{\Pi}, N) &= \frac{N\Pi}{2} \exp\left(-\frac{N}{4}(\Pi^2 + \hat{\Pi}^2)\right) I_0\left(\frac{N\Pi\hat{\Pi}}{2}\right) \\
&= \frac{N\Pi}{2} \exp\left(-\frac{N(\Pi - \hat{\Pi})^2}{4}\right) i_0\left(\frac{N\Pi\hat{\Pi}}{2}\right),
\end{aligned}
\tag{38}
$$

where I_0 is the modified Bessel function of order zero, $i_0(x) = \exp(-|x|)I_0(x)$, and N is related to the statistical quality, e.g., the number of photons detected in a photon counting experiment. Use of the scaled modified Bessel function $i_0(x)$ is advantageous for numerical implementation. Expectation value and standard deviation of Π are then

$$
\bar{\Pi} = \sqrt{\frac{\pi}{16N}} \exp\left(-\frac{N\hat{\Pi}^2}{8}\right) \left[(4 + N\hat{\Pi}^2)I_0\left(\frac{N\hat{\Pi}^2}{8}\right) + N\hat{\Pi}^2 I_1\left(\frac{N\hat{\Pi}^2}{8}\right)\right],
\tag{39}
$$

$$
\bar{\sigma}_\Pi = \left(\hat{\Pi}^2 + \frac{4}{N} - \bar{\Pi}^2\right)^{1/2},
\tag{40}
$$

where I_1 is the modified Bessel function of order 1. For each polarization measurement $\Pi_m \pm \sigma_m$, we determine N by numerically solving $\bar{\sigma}_\Pi = \sigma_m$ for N assuming $\hat{\Pi} = \Pi_m$. This allows us to calculate the cumulative probability by numeric integration of Equation (38):

$$
P(\Pi \leq \Pi_{\max}|\Pi_m, N) = \int_0^{\Pi_{\max}} P(\Pi|\Pi_m, N)d\Pi.
\tag{41}
$$

This result quantifies the probability that a given set of SME coefficients is compatible with the measurement. Typical examples of the integral as a function of the upper limit $\Pi_{\max}$ are shown in Figure 11. In practice, we replace $P(\Pi|\hat{\Pi}, N)$ from Equation (38) with a simple Gaussian distribution with mean Π_m and standard deviation σ_m if $\Pi_m/\sigma_m > 10$ due to numerical issues when evaluating the Bessel functions for large values of N. The error in this case is $P_{\mathrm{Gauss}}(x < 0|\mu > 10\sigma) < 7.7 \times 10^{-24}$. An example of this cumulative distribution as a function of one of the SME coefficients is shown in Figure 12.

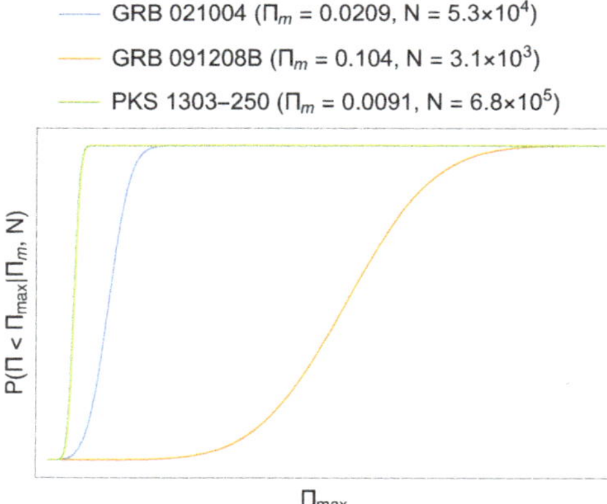

Figure 11. Three typical examples of the integral in Equation (41) as function of the upper limit $\Pi_{\max}$. The three cases illustrate a range of measured polarization fractions Π_m and uncertainties σ_m.

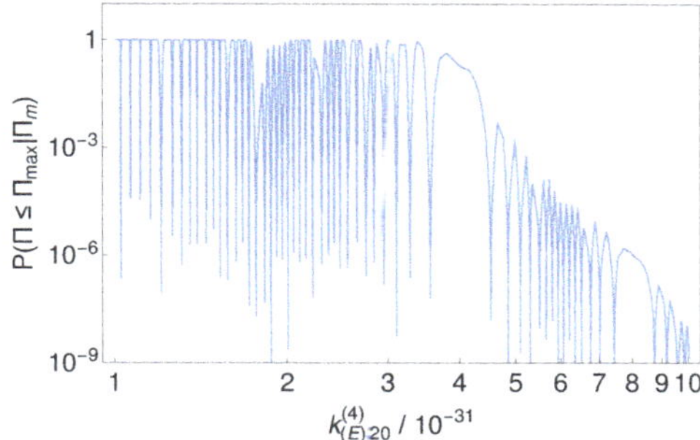

Figure 12. Probability that the observed polarization fraction of GRB 091208B is smaller than the maximum theoretically possible polarization fraction according to Equation (41) (see Figure 10) as function of the SME coefficient $k^{(4)}_{(E)20}$ when keeping all other coefficients at 0. The measured polarization fraction is 0.104(25).

We applied this method to a large sample of polarization measurements of AGN [38,39]. From this catalog, we selected 36 sources for which polarization was measured with at least 5σ significance. Furthermore, we included optical polarization measurements of eight GRBs [40–51]. This selection of measurements is identical to the data used in Ref. [20] with the exception that we were unable to use data from GRB 090102 because no polarization angle was published [52]. Details about all astrophysical sources and measurements used here are shown in Tables A1 and A2 in Appendix A.

4. Results

The goal of this work is to obtain constraints on the individual coefficients $k^{(4)}_{(E)2m}$ and $k^{(4)}_{(B)2m}$. From the measurements described in the previous section, we find the combined probability distribution

$$P(k^{(4)}_{(E)2m}, k^{(4)}_{(B)2m}) = \prod_{\text{measurements } i} P_i(k^{(4)}_{(E)2m}, k^{(4)}_{(B)2m}), \tag{42}$$

where the $P_i(k^{(4)}_{(E)2m}, k(B)2m^{(4)})$ are the probabilities that each measurement is compatible with the given set of SME coefficients calculated as described in the previous section (Equations (29) and (41)). The product goes over all measurements of both types, listed in Tables A1–A3 in Appendix A.

To calculate constraints on each SME coefficient, we evaluate the probability in Equation (42) and marginalize over all other coefficients. Using the Metropolis-Hastings Markov Chain Monte Carlo method [53] with a normal distribution with $\sigma = 6 \times 10^{-35}$ for each coefficient as the proposal distribution $g(\delta k^{(4)}_{(E)2m}, \delta k^{(4)}_{(B)2m})$, we sample the combined distribution. At each step t, we randomly choose a new set of parameters $k^{(4)}_{(E,B)2m,t+1} = k^{(4)}_{(E,B)2m,t} + \delta k^{(4)}_{(E,B)2m}$ according to g and calculate the acceptance ratio $\alpha = P(k^{(4)}_{(E)2m,t+1}, k^{(4)}_{(B)2m,t+1})/P(k^{(4)}_{(E)2m,t}, k^{(4)}_{(B)2m,t})$ according to Equation (42). The step is accepted if $u \leq \alpha$ for a uniform random number $u \in [0,1]$, otherwise we set $k^{(4)}_{(E,B)2m,t+1} = k^{(4)}_{(E,B)2m,t}$. In this way, we iteratively construct a set of coefficients $k^{(4)}_{(E,B)2m}$ that can be shown to be distributed according to the probability density in Equation (42).

The choice of proposal distribution g leads to a step acceptance rate of about 13 %. Starting from all coefficients set to 0, we discarded the first 10,000 steps to reduce the dependence of the result on the initial set of coefficients, and then took 10^7 steps recording each set of selected SME coefficients. The distribution of these sets of coefficients is proportional to $P(k^{(4)}_{(E)2m}, k^{(4)}_{(B)2m})$. From the distribution of values of each individual coefficient, we then find the 5th and 95th percentile as lower and upper limits. The resulting distributions are shown in Figure A1 in Appendix B and the corresponding upper and lower limits on the SME coefficients are listed in Table 2. The constraints turned out to be symmetrical around 0 within at least two-digit precision.

We also derived constraints from each of the two methods individually as a cross check. When using only the spectrally integrated measurements described in Section 3.2, the upper limits on the coefficients are about a factor 100 larger than those in Table 2. Consistently, we found that the results obtained when only using the spectropolarimetric results from Section 3.1 differ from the final ones on the order of 1 %.

By combining multiple spectropolarimetric measurements, the probability spikes in Figure 5 are eliminated as their location depends on source distance, direction, and measured polarization angle. The "ringing" observed in the probability distribution derived from each spectrally integrated measurement seen in Figure 12 cancels to some degree when combining multiple spectrally integrated measurements. Furthermore, the spectropolarimetric measurements are significantly more constraining than the spectrally integrated results. Combined, these effects result in the relatively smooth probability distribution of each SME coefficient shown in Figure A1.

Figure A2 in Appendix B shows the correlation between the different coefficients. There is significant correlation among most pairs of coefficients most likely due to the non-uniform sky coverage. Therefore, the best way to improve on these results is to improve the sky coverage particularly with spectropolarimetric measurements, which at this point were not available for most of the Southern sky.

Table 2. Limits at the 95 % confidence level on all independent SME parameters $k^{(4)}_{(E)2m}$ and $k^{(4)}_{(B)2m}$ obtained in this analysis. The dependent parameters $k^{(4)}_{(E)2(-m)}$ and $k^{(4)}_{(B)2(-m)}$ can be calculated according to Equation (6).

$$|k^{(4)}_{(E)20}| < 2.4 \times 10^{-34}$$
$$|\operatorname{Re}(k^{(4)}_{(E)21})| < 1.0 \times 10^{-34}$$
$$|\operatorname{Im}(k^{(4)}_{(E)21})| < 1.6 \times 10^{-34}$$
$$|\operatorname{Re}(k^{(4)}_{(E)22})| < 2.4 \times 10^{-34}$$
$$|\operatorname{Im}(k^{(4)}_{(E)22})| < 2.6 \times 10^{-34}$$
$$|k^{(4)}_{(B)20}| < 1.5 \times 10^{-34}$$
$$|\operatorname{Re}(k^{(4)}_{(B)21})| < 2.2 \times 10^{-34}$$
$$|\operatorname{Im}(k^{(4)}_{(B)21})| < 1.4 \times 10^{-34}$$
$$|\operatorname{Re}(k^{(4)}_{(B)22})| < 1.9 \times 10^{-34}$$
$$|\operatorname{Im}(k^{(4)}_{(B)22})| < 2.5 \times 10^{-34}$$

5. Discussion

Using optical polarization measurements of 63 AGN and GRBs, we searched for signals of Lorentz invariance violation. We derived 95 % confidence level limits on each of the 10 coefficients with mass dimension $d = 4$ of the minimal SME on the order of 10^{-34}, as listed in Table 2. The results summarized in Table 2 are the first constraints on the individual SME parameters in this sector. Furthermore, the results are based on highly significant polarization measurements with well quantified uncertainties. In comparison, the limits published in [19] only constrain combinations of SME parameters, and are based on measurements of the polarization of the X-ray emission from GRBs with large systematic uncertainties.

While there may not be a theory of quantum gravity currently under active consideration that predicts a form of Lorentz invariance violation described by these coefficients, there is no generally accepted quantum gravity theory, either. A broad search for possible effects is, thus, warranted as neither the form of potential Lorentz invariance violation, nor its degree is known a priori. Given the tight constraints on the SME coefficients of $d = 4$ presented here and in prior work [4], any significant violation of Lorentz symmetry at the Planck scale must be strongly suppressed at lower energies. If a hierarchy of scales exists, suppression with $(M_{low}/M_{Planck})^n$ is possible, where M_{low} is the lower mass scale and n is some power. SUSY-breaking is one scenario that could provide such a hierarchy [54]. Given extensive direct searches for deviations from the Standard Model of particle physics at energies up to about 1 TeV [55], $M_{low} \gg 1\,\text{TeV}$ must be assumed. A model with $M_{low} = 10\,\text{TeV}$ and $n = 2$ would lead to a suppression of the order of 10^{-30}. In this scenario, Lorentz invariance violating effects of order unity at the Planck scale due to operators of mass dimension $d = 4$ can be ruled out by the constraints presented here. For a similar model to be viable and to result in significant Lorentz invariance violating effects at the Planck scale, one must assume that either $n > 2$ or $M_{low} \gg 10\,\text{TeV}$.

While optical polarization measurements are a powerful tool to constrain coefficients of $d = 4$ and $d = 5$ [20], at higher mass dimension polarization measurements at higher energy will be significantly stronger due to the E^{d-3} dependence of the polarization signature. The Imaging X-ray Polarimetry Explorer (IXPE, [56]) to be launched in early 2021 will provide highly significant X-ray polarization measurements of a large number of astrophysical objects and one can expect that a similar analysis with those data will significantly improve on the results presented here. Future Compton gamma-ray telescopes, such as the proposed All-sky Medium-Energy Gamma-ray Observatory (AMEGO [57]) or the Compton Spectrometer and Imager (COSI [58]) will be sensitive to gamma-ray polarization up to MeV energies. These future instruments will significantly enhance our capability to search for Lorentz invariance violation in the photon sector.

Funding: This research received funding from NASA Grant NNX14AD19G and DOE Grant DE-FG02-91ER40628.

Acknowledgments: I would like to thank Henric Krawczynski, Alan Kostelecký, Matthew Mewes, Stefano Covino, David Mattingly, Jim Buckley, Floyd Stecker, Paul Smith, and Manel Errando for fruitful discussions. Part of this work was done while I was at Washington University in St. Louis, and I would like to thank the McDonnell Center for the Space Sciences at Washington University in St. Louis for support. Data from the Steward Observatory spectropolarimetric monitoring project were used, which is supported by Fermi Guest Investigator grants NNX08AW56G, NNX09AU10G, NNX12AO93G, and NNX15AU81G. Furthermore, I made use of the SIMBAD database, operated at CDS, Strasbourg, France.

Conflicts of Interest: The author declares no conflict of interest.

Appendix A. Astrophysical Sources

Tables A1–A3 list all astrophysical sources used in this analysis, where the spectropolarimetric results are given in Table A3.

Table A1. Sources observed by Sluse et al. [38,39], including the coordinates and redshifts listed in the reference. All observations were made using the EFOSC2 instrument.

Source	RA J2000 [°]	Dec. J2000 [°]	Redshift z	Polarization [%]	Position Angle [°]
SDSS J0242+0049	40.591	+0.820	2.071	1.47(24)	−13(5)
FIRST03133+0036	48.328	+0.606	1.250	1.48(29)	−55(6)
FIRST J0809+2753	122.256	+27.895	1.511	1.75(20)	73(3)
PG 0946+301	147.421	+29.922	1.220	1.65(19)	−66(3)
PKS 1124-186	171.768	−17.045	1.048	11.68(36)	37(1)
He 1127-1304	172.583	−12.653	0.634	1.32(13)	46(3)
2QZ J114954+0012	177.479	+0.215	1.596	1.57(22)	−24(4)
SDSS J1206+0023	181.615	+0.393	2.331	0.94(15)	−57(5)
SDSS J1214-0001	183.673	+0.027	1.041	2.40(32)	−77(4)
PKS 1219+04	185.594	+4.221	0.965	5.56(15)	−61(1)
PKS 1222+037	186.218	+3.514	0.960	2.51(22)	−82(2)
TON 1530	186.364	+22.587	2.058	0.92(14)	−11(4)

Table A1. *Cont.*

Source	RA J2000 [°]	Dec. J2000 [°]	Redshift z	Polarization [%]	Position Angle [°]
SDSS J1234+0057	188.616	+0.966	1.532	1.35(23)	2(5)
PG 1254+047	194.250	+4.459	1.018	0.84(15)	−4(5)
PKS 1256-229	194.785	−22.823	1.365	22.32(15)	−23(1)
SDSS J1302-0037	195.534	+0.626	1.672	1.37(20)	35(4)
PKS 1303-250	196.564	−24.711	0.738	0.91(17)	−75(5)
FIRST J1312+2319	198.056	+23.333	1.508	1.10(16)	−14(4)
SDSS J1323-0038	200.769	+0.649	1.827	1.13(21)	15(5)
CTS J13.07	205.518	−17.700	2.210	0.83(15)	20(5)
SDSS J1409+0048	212.328	+0.807	1.999	3.91(28)	30(2)
HS 1417+2547	215.055	+25.568	2.200	1.03(18)	−66(5)
FIRST J1427+2709	216.765	+27.161	1.170	1.35(25)	80(5)
FIRST J21079-0620	316.990	−5.664	0.644	1.12(22)	−33(6)
SDSS J2131-0700	322.912	−6.996	2.048	1.78(32)	44(5)
PKS 2204-54	331.932	−52.224	1.206	1.81(26)	−50(4)
PKS 2227-445	337.735	−43.725	1.326	5.26(48)	18(3)
PKS 2240-260	340.860	−24.258	0.774	14.78(21)	−49(1)
PKS 2301+06	346.118	+6.336	1.268	3.69(26)	−17(2)
SDSS J2319-0024	349.995	+0.414	1.889	1.85(30)	−16(5)
PKS 2320-035	350.883	−2.715	1.411	9.56(20)	90(1)
PKS 2332-017	353.835	−0.481	1.184	4.86(19)	−88(1)
PKS 2335-027	354.489	−1.484	1.072	3.55(30)	−70(2)
SDSS J2352+0105	358.159	+1.098	2.156	1.59(26)	27(5)
SDSS J2356-0036	359.120	+0.601	2.936	1.81(34)	16(5)
QSO J2359-12	359.973	−11.303	0.868	4.12(20)	−29(1)

Table A2. Optical GRB measurements.

Source	Instrument	RA J2000 [°]	Dec. J2000 [°]	Redshift z	Polarization [%]	Position Angle [°]	Refs.
GRB 990510	FORS1 R-band	204.532	−80.497	1.619	1.6 ± 0.2	−84 ± 4	[41–43]
GRB 990712	FORS1 R-band	337.971	−73.408	0.430	2.9 ± 0.4	−59 ± 4	[44]
GRB 020813	FORS1 V-band	296.674	−19.601	1.25	1.42 ± 0.25	−43 ± 4	[47,48]
GRB 021004	NOT/ALFOSC	6.728	+18.928	2.330	2.1 ± 0.6	−7 ± 8	[45]
GRB 030329	NOT/ALFOSC	161.208	+21.522	0.169	2.4 ± 0.4	65 ± 7	[46]
GRB 091018	FORS2	32.192	−57.55	0.97	3.25 ± 0.35	57 ± 6	[49]
GRB 091208B	HOWPol	29.410	16.881	1.06	10.4 ± 2.5	−88 ± 6	[50]
GRB 121024A	FORS2	70.467	−12.268	2.298	4.83 ± 0.20	173	[51] †

† No uncertainty of the position angle was specified in the paper. We are able to use this measurement because we always find the position angle the source to exactly match the observed position angle, irrespective of its uncertainty.

Table A3. Sources selected from the Steward Observatory spectropolarimetric monitoring project [26]. The second column lists the highest observed polarization fraction during cycles 1–7 of the monitoring program. Coordinates have been obtained from the SIMBAD database [59]. Individual references are given for the red shifts. The last two columns give the position angle at the median energy of the linear fit and the differential change in position angle.

Source	P_{max} [%]	RA J2000 [°]	Dec. J2000 [°]	Redshift z		PA at 2.26 eV [°]	ρ [°/eV]
3C 454.3	18.83	343.491	+16.148	0.859	[60]	61.19(23)	−0.3(10)
4C 14.23	20.32	111.320	+14.420	1.038	[61]	−34.42(18)	−0.7(7)
4C 28.07	30.30	39.468	+28.802	1.206	[61]	−62.36(9)	0.46(32)
AO 0235+164	39.79	39.662	+16.616	0.940	[62]	−20.25(5)	0.16(17)

Table A3. *Cont.*

Source	P_{max} [%]	RA J2000 [°]	Dec. J2000 [°]	Redshift z		PA at 2.26 eV [°]	ρ [°/eV]
B2 1633+382	27.26	248.815	+38.135	1.813	[63]	−3.71(7)	−0.69(26)
B2 1846+32A	28.88	282.092	+32.317	0.800	[61]	2.39(5)	1.49(18)
B3 0650+453	16.16	103.599	+45.240	0.928	[61]	79.3(5)	3.4(18)
B3 1343+451	10.07	206.388	+44.883	2.534	[61]	35.6(6)	−0.7(26)
BZU J0742+5444	21.73	115.666	+54.740	0.723	[64]	−87.84(19)	2.2(6)
CTA 26	26.21	54.879	−1.777	0.852	[65]	67.49(7)	−0.13(24)
CTA 102	23.97	338.152	+11.731	1.037	[66]	64.81(9)	1.29(34)
MG1 J123931+0443	33.61	189.886	+4.718	1.760	[63]	−72.28(8)	−0.02(32)
OJ 248	18.09	127.717	+24.183	0.941	[63]	−79.23(11)	−0.9(4)
PKS 0420-014	28.67	65.816	−1.343	0.916	[67]	10.23(9)	0.44(31)
PKS 0454-234	35.27	74.263	−23.414	1.003	[60]	2.49(6)	0.22(20)
PKS 0502+049	17.59	76.347	+4.995	0.954	[68]	−87.82(13)	2.1(5)
PKS 0805-077	28.27	122.065	−7.853	1.837	[60]	41.7(4)	−1.4(11)
PKS 1118-056	22.54	170.355	−5.899	1.297	[60]	37.26(9)	1.4(7)
PKS 1124-186	10.49	171.768	−18.955	1.048	[62]	−83.18(19)	0.2(7)
PKS 1244-255	13.97	191.695	−25.797	0.638	[60]	−14.07(11)	1.9(4)
PKS 1441+252	37.70	220.987	+25.029	0.939	[61]	−72.34(8)	0.51(28)
PKS 1502+106	45.16	226.104	+10.494	1.839	[63]	65.43(12)	0.1(4)
PKS 2032+107	12.36	308.843	+10.935	0.601	[61]	68.4(10)	9.2(34)
PMN J2345-1555	32.69	356.302	−15.919	0.621	[61]	−0.46(4)	1.30(15)
S4 1030+61	37.71	158.464	+60.852	1.400	[63]	−59.69(12)	−1.3(4)
SDSS J084411+5312	18.72	131.049	+53.214	3.704	[63]	8.4(4)	−2.0(17)
Ton 599	33.16	179.883	+29.246	0.724	[63]	−52.69(7)	0.86(26)

Appendix B. Coefficient Distributions and Correlations

Figure A1 shows the probability distributions for all 10 SME coefficients with $d = 4$ derived in this analysis. Figure A2 shows the correlations between coefficients.

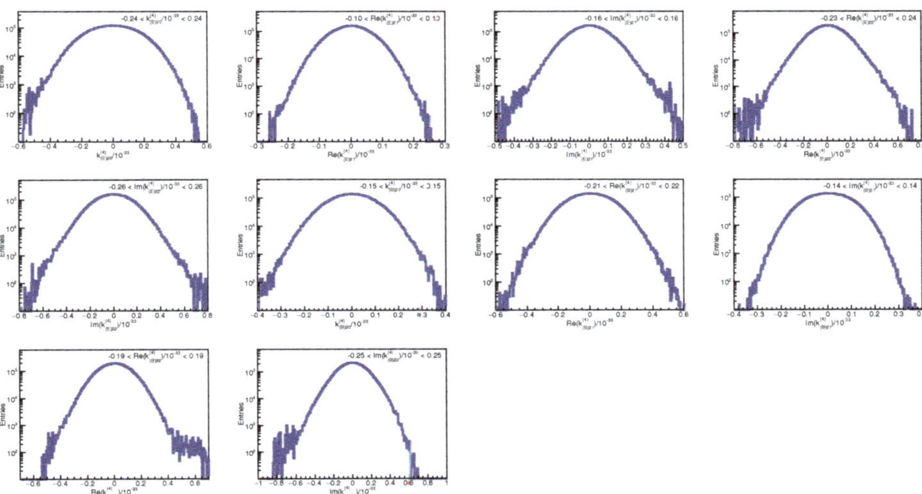

Figure A1. Distributions of all coefficients derived using the MCMC sampling of the coefficients space, marginalized over the nine other coefficients. The constraints listed in each panel are 5th–95th percentile.

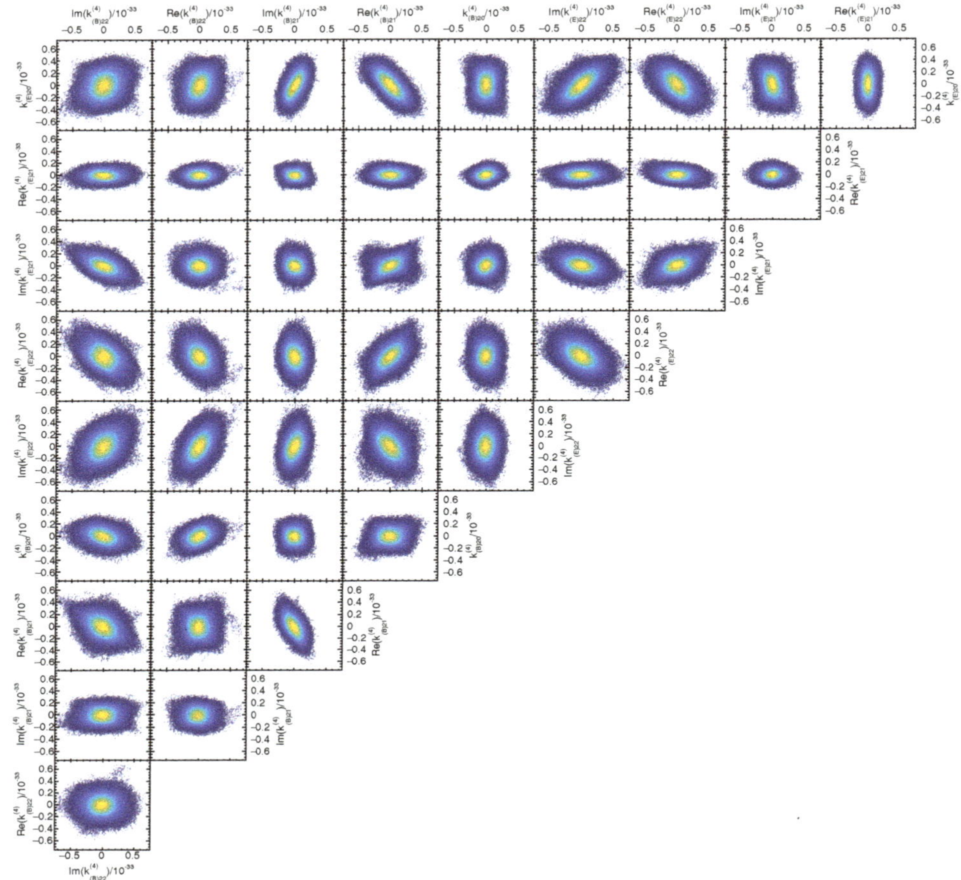

Figure A2. Correlations between SME coefficients found this analysis.

References

1. Michelson, A.A.; Morley, E.W. On the Relative Motion of the Earth and the Luminiferous Aether. *Phil. Mag.* **1887**, *24*, 449. [CrossRef]
2. Kennedy, R.J.; Thorndike, E.M. Experimental Establishment of the Relativity of Time. *Phys. Rev.* **1932**, *42*, 400–418. [CrossRef]
3. Ives, H.E.; Stilwell, G.R. An Experimental study of the rate of a moving atomic clock. *J. Opt. Soc. Am.* **1938**, *28*, 215. [CrossRef]
4. Kostelecký, V.A.; Russell, N. Data Tables for Lorentz and CPT Violation. *Rev. Mod. Phys.* **2011**, *83*, 11–31. [updated at arXiv:hep-ph/0801.0287v11]. [CrossRef]
5. Amelino-Camelia, G.; Ellis, J.; Mavromatos, N.E.; Nanopoulos, D.V.; Sarkar, S. Tests of quantum gravity from observations of γ-ray bursts. *Nature* **1998**, *393*, 763–765. [CrossRef]
6. Garay, L.J. Spacetime Foam as a Quantum Termal Bath. *Phys. Rev. Lett.* **1998**, *80*, 2508–2511. [CrossRef]
7. Gambini, R.; Pullin, J. Nonstandard optics from quantum space-time. *Phys. Rev. D* **1999**, *59*, 124021. [CrossRef]
8. Mattingly, D. Modern Tests of Lorentz Invariance. *Living Rev. Relativity* **2005**, *8*, 5. Available online: http://www.livingreviews.org/lrr-2005-5 (accessed on 31 August 2018). [CrossRef] [PubMed]

9. Jacobson, T.; Liberati, S.; Mattingly, D. Lorentz violation at high energy: Concepts, phenomena, and astrophysical constraints. *Ann. Phys.* **2006**, *321*, 150–196. [CrossRef]

10. Liberati, S.; Maccione, L. Lorentz Violation: Motivation and New Constraints. *Annu. Rev. Nucl. Part. Sci.* **2009**, *59*, 245–267. [CrossRef]

11. Kostelecký, V.A.; Mewes, M. Astrophysical Tests of Lorentz and CPT Violation with Photons. *Astrophys. J. Lett.* **2008**, *689*, L1–L4. [CrossRef]

12. Kostelecký, V.A.; Mewes, M. Electrodynamics with Lorentz-violating operators of arbitrary dimension. *Phys. Rev. D* **2009**, *80*, 015020. [CrossRef]

13. Abdo, A.A.; Ackermann, M.; Ajello, M.; Asano, K.; Atwood, W.B.; Axelsson, M.; Baldini, L.; Ballet, J.; Barbiellini, G.; Baring, M.G.; et al. A limit on the variation of the speed of light arising from quantum gravity effects. *Nature* **2009**, *462*, 331–334. [CrossRef] [PubMed]

14. Vasileiou, V.; Jacholkowska, A.; Piron, F.; Bolmont, J.; Couturier, C.; Granot, J.; Stecker, F.W.; Cohen-Tanugi, J.; Longo, F. Constraints on Lorentz invariance violation from Fermi-Large Area Telescope observations of gamma-ray bursts. *Phys. Rev. D* **2013**, *87*, 122001. [CrossRef]

15. Abramowski, A.; Aharonian, F.; Ait Benkhali, F.; Akhperjanian, A.G.; Angüner, E.O.; Backes, M.; Balenderan, S.; Balzer, A.; Barnacka, A.; Becherini, Y.; et al. The 2012 Flare of PG 1553+113 Seen with H.E.S.S. and Fermi-LAT. *Astrophys. J.* **2015**, *802*, 65. [CrossRef]

16. Kislat, F.; Krawczynski, H. Search for anisotropic Lorentz invariance violation with γ-rays. *Phys. Rev. D* **2015**, *92*, 045016. [CrossRef]

17. Wei, J.J.; Wu, X.F.; Zhang, B.B.; Shao, L.; Mészáros, P.; Kostelecký, V.A. Constraining Anisotropic Lorentz Violation via the Spectral-lag Transition of GRB 160625B. *Astrophys. J.* **2017**, *842*, 115. [CrossRef]

18. Kaaret, P. X-ray clues to viability of loop quantum gravity. *Nature* **2004**, *427*, 287. [CrossRef] [PubMed]

19. Kostelecký, V.A.; Mewes, M. Constraints on Relativity Violations from Gamma-Ray Bursts. *Phys. Rev. Lett.* **2013**, *110*, 201601. [CrossRef] [PubMed]

20. Kislat, F.; Krawczynski, H. Planck-scale constraints on anisotropic Lorentz and CPT invariance violations from optical polarization measurements. *Phys. Rev. D* **2017**, *95*, 083013. [CrossRef]

21. Komatsu, E.; Smith, K.M.; Dunkley, J.; Bennett, C.L.; Gold, B.; Hinshaw, G.; Jarosik, N.; Larson, D.; Nolta, M.R.; Page, L.; et al. Seven-year Wilkinson Microwave Anisotropy Probe (WMAP) Observations: Cosmological Interpretation. *Astrophys. J. Suppl. Ser.* **2011**, *192*, 18. [CrossRef]

22. Brown, M.L.; Ade, P.; Bock, J.; Bowden, M.; Cahill, G.; Castro, P.G.; Church, S.; Culverhouse, T.; Friedman, R.B.; Ganga, K.; et al. Improved Measurements of the Temperature and Polarization of the Cosmic Microwave Background from QUaD. *Astrophys. J.* **2009**, *705*, 978–999. [CrossRef]

23. Pagano, L.; de Bernardis, P.; de Troia, G.; Gubitosi, G.; Masi, S.; Melchiorri, A.; Natoli, P.; Piacentini, F.; Polenta, G. CMB polarization systematics, cosmological birefringence, and the gravitational waves background. *Phys. Rev. D* **2009**, *80*, 043522. [CrossRef]

24. Kostelecký, V.A.; Melissinos, A.C.; Mewes, M. Searching for photon-sector Lorentz violation using gravitational-wave detectors. *Phys. Lett. B* **2016**, *761*, 1–7. [CrossRef]

25. Contopoulos, G.; Jappel, A. Transactions of the International Astronomical Union, Volume XVB. In *Proceedings of the Fifteenth General Assembly, Sydney 1973 and Extraordinary Assembly, Poland 1973*; Springer Science & Business Media: Berlin, Germany, 1980.

26. Smith, P.S.; Montiel, E.; Rightley, S.; Turner, J.; Schmidt, G.D.; Jannuzi, B.T. Coordinated Fermi/Optical Monitoring of Blazars and the Great 2009 September Gamma-ray Flare of 3C 454.3. Available online: http://arxiv.org/abs/0912.3621 (accessed on 6 September 2018).

27. SVO Filter Profile Service. Available online: http://svo2.cab.inta-csic.es/svo/theory/fps3/index.php?&mode=browse&gname=NOT&gname2=ALFOSC&all=0 (accessed on 23 November 2015).

28. O'Brien, K.; Marconi, G.; Dumas, C. *Very Large Telescope Paranal Science Operations FORS User Manual*; Technical Report VLT-MAN-ESO-13100-1543, Version 82.1; European Southern Observatory: Garching, Germany, 2008.

29. Boffin, H.; Dumas, C.; Kaufer, A. *Very Large Telescope Paranal Science Operations FORS2 User Manual*; Technical Report VLT-MAN-ESO-13100-1543, Version 96.0; European Southern Observatory: Garching, Germany, 2015.

30. Hau, G.K.T.; Patat, F. *EFOSC2 User's Manual*; Technical Report LSO-MAN-ESO-36100-0004, Version 2.0; European Southern Observatory: Garching, Germany, 2003.

31. Kawabata, K.S.; Nagae, O.; Chiyonobu, S.; Tanaka, H.; Nakaya, H.; Suzuki, M.; Kamata, Y.; Miyazaki, S.; Hiragi, K.; Miyamoto, H.; et al. Wide-field one-shot optical polarimeter: HOWPol. In *Ground-Based and Airborne Instrumentation for Astronomy II*; SPIE: Marseille, France, 2008; Volume 7014, p. 70144L.

32. Marscher, A.P. Turbulent, Extreme Multi-zone Model for Simulating Flux and Polarization Variability in Blazars. *Astrophys. J.* **2014**, *780*, 87. [CrossRef]

33. Zhang, H.; Li, H.; Guo, F.; Taylor, G. Polarization Signatures of Kink Instabilities in the Blazar Emission Region from Relativistic Magnetohydrodynamic Simulations. *Astrophys. J.* **2017**, *835*, 125. [CrossRef]

34. Scarpa, R.; Falomo, R. Are high polarization quasars and BL Lacertae objects really different? A study of the optical spectral properties. *Astron. Astrophys.* **1997**, *325*, 109–123.

35. Deng, W.; Zhang, B.; Li, H.; Stone, J.M. Magnetized Reverse Shock: Density-fluctuation-induced Field Distortion, Polarization Degree Reduction, and Application to GRBs. *Astrophys. J. Lett.* **2017**, *845*, L3. [CrossRef]

36. Weisskopf, M.C.; Elsner, R.F.; Kaspi, V.M.; O'Dell, S.L.; Pavlov, G.G.; Ramsey, B.D. X-Ray Polarimetry and Its Potential Use for Understanding Neutron Stars. In *Astrophysics and Space Science Library*; Becker, W., Ed.; Springer: Berlin, Germany, 2009; Volume 357, p. 589.

37. Krawczynski, H. Analysis of the data from Compton X-ray polarimeters which measure the azimuthal and polar scattering angles. *Astropart. Phys.* **2011**, *34*, 784–788. [CrossRef]

38. Sluse, D.; Hutsemékers, D.; Lamy, H.; Cabanac, R.; Quintana, H. New optical polarization measurements of quasi-stellar objects. The data. *Astron. Astrophys.* **2005**, *433*, 757–764. [CrossRef]

39. Sluse, D.; Hutsemékers, D.; Lamy, H.; Cabanac, R.; Quintana, H. Optical Polarization of 203 QSOs (Sluse+, 2005). Available online: http://vizier.u-strasbg.fr/viz-bin/VizieR?-source=J/A+A/433/757 (accessed on 6 September 2018).

40. Covino, S.; Götz, D. Polarization of prompt and afterglow emission of Gamma-Ray Bursts. *Astron. Astrophys. Trans.* **2016**, *29*, 205–244.

41. Wijers, R.A.M.J.; Vreeswijk, P.M.; Galama, T.J.; Rol, E.; van Paradijs, J.; Kouveliotou, C.; Giblin, T.; Masetti, N.; Palazzi, E.; Pian, E.; et al. Detection of Polarization in the Afterglow of GRB 990510 with the ESO Very Large Telescope. *Astrophys. J.* **1999**, *523*, L33–L36. [CrossRef]

42. Covino, S.; Lazzati, D.; Ghisellini, G.; Saracco, P.; Campana, S.; Chincarini, G.; di Serego, S.; Cimatti, A.; Vanzi, L.; Pasquini, L.; et al. GRB 990510: Linearly polarized radiation from a fireball. *Astron. Astrophys.* **1999**, *348*, L1–L4.

43. Covino, S.; Lazzati, D.; Ghisellini, G.; Saracco, P.; Campana, S.; Chincarini, G.; di Serego Alighieri, S.; Cimatti, A.; Vanzi, L.; Pasquini, L.; et al. GRB 990510. *IAU Circ.* **1999**, *7172*, 3.

44. Rol, E.; Wijers, R.A.M.J.; Vreeswijk, P.M.; Kaper, L.; Galama, T.J.; van Paradijs, J.; Kouveliotou, C.; Masetti, N.; Pian, E.; Palazzi, E.; et al. GRB 990712: First Indication of Polarization Variability in a Gamma-Ray Burst Afterglow. *Astrophys. J.* **2000**, *544*, 707–711. [CrossRef]

45. Rol, E.; Wijers, R.A.M.J.; Fynbo, J.P.U.; Hjorth, J.; Gorosabel, J.; Egholm, M.P.; Castro Cerón, J.M.; Castro-Tirado, A.J.; Kaper, L.; Masetti, N.; et al. Variable polarization in the optical afterglow of GRB 021004. *Astron. Astrophys.* **2003**, *405*, L23–L27. [CrossRef]

46. Greiner, J.; Klose, S.; Reinsch, K.; Martin Schmid, H.; Sari, R.; Hartmann, D.H.; Kouveliotou, C.; Rau, A.; Palazzi, E.; Straubmeier, C.; et al. Evolution of the polarization of the optical afterglow of the γ-ray burst GRB030329. *Nature* **2003**, *426*, 157–159. [CrossRef] [PubMed]

47. Gorosabel, J.; Rol, E.; Covino, S.; Castro-Tirado, A.J.; Castro Cerón, J.M.; Lazzati, D.; Hjorth, J.; Malesani, D.; Della Valle, M.; di Serego Alighieri, S.; et al. GRB 020813: Polarization in the case of a smooth optical decay. *Astron. Astrophys.* **2004**, *422*, 113–119. [CrossRef]

48. Barth, A.J.; Sari, R.; Cohen, M.H.; Goodrich, R.W.; Price, P.A.; Fox, D.W.; Bloom, J.S.; Soderberg, A.M.; Kulkarni, S.R. Optical Spectropolarimetry of the GRB 020813 Afterglow. *Astrophys. J.* **2003**, *584*, L47–L51. [CrossRef]

49. Wiersema, K.; Curran, P.A.; Krühler, T.; Melandri, A.; Rol, E.; Starling, R.L.C.; Tanvir, N.R.; van der Horst, A.J.; Covino, S.; Fynbo, J.P.U.; et al. Detailed optical and near-infrared polarimetry, spectroscopy and broad-band photometry of the afterglow of GRB 091018: Polarization evolution. *Mon. Not. Royal Astron. Soc.* **2012**, *426*, 2–22. [CrossRef]

50. Uehara, T.; Toma, K.; Kawabata, K.S.; Chiyonobu, S.; Fukazawa, Y.; Ikejiri, Y.; Inoue, T.; Itoh, R.; Komatsu, T.; Miyamoto, H.; et al. GRB 091208B: First Detection of the Optical Polarization in Early Forward Shock Emission of a Gamma-Ray Burst Afterglow. *Astrophys. J.* **2012**, *752*, L6. [CrossRef]

51. Wiersema, K.; Covino, S.; Toma, K.; van der Horst, A.J.; Varela, K.; Min, M.; Greiner, J.; Starling, R.L.C.; Tanvir, N.R.; Wijers, R.A.M.J.; et al. Circular polarization in the optical afterglow of GRB 121024A. *Nature* **2014**, *509*, 201–204. [CrossRef] [PubMed]

52. Steele, I.A.; Mundell, C.G.; Smith, R.J.; Kobayashi, S.; Guidorzi, C. Ten per cent polarized optical emission from GRB090102. *Nature* **2009**, *462*, 767–769. [CrossRef] [PubMed]

53. Hastings, W.K. Monte Carlo Sampling Methods using Markov Chains and their Applications. *Biometrika* **1970**, *57*, 97–109. [CrossRef]

54. Bolokhov, P.A.; Groot Nibbelink, S.; Pospelov, M. Lorentz violating supersymmetric quantum electrodynamics. *Phys. Rev. D* **2005**, *72*, 015013. [CrossRef]

55. Tanabashi, M.; Hagiwara, K.; Hikasa, K.; Nakamura, K.; Sumino, Y.; Takahashi, F.; Tanaka, J.; Agashe, K.; Aielli, G.; Amsler, C.; et al. Review of Particle Physics. *Phys. Rev. D* **2018**, *98*, 030001. [CrossRef]

56. Weisskopf, M.C.; Ramsey, B.; O'Dell, S.L.; Tennant, A.; Elsner, R.; Soffita, P.; Bellazzini, R.; Costa, E.; Kolodziejczak, J.; Kaspi, V.; et al. The Imaging X-ray Polarimetry Explorer (IXPE). *Results Phys.* **2016**, *6*, 1179–1180. [CrossRef]

57. Moiseev, A. All-Sky Medium Energy Gamma-ray Observatory (AMEGO). *Proc. Sci.* **2018**, *ICRC2017*, 798.

58. Yang, C.Y.; Lowell, A.; Zoglauer, A.; Tomsick, J.; Chiu, J.L.; Kierans, C.; Sleator, C.; Boggs, S.; Chang, H.K.; Jean, P.; et al. The polarimetric performance of the Compton Spectrometer and Imager (COSI). In *Society of Photo-Optical Instrumentation Engineers (SPIE) Conference Series*; SPIE: Austin, TX, USA, 2018; Volume 10699, p. 106992K.

59. Wenger, M.; Ochsenbein, F.; Egret, D.; Dubois, P.; Bonnarel, F.; Borde, S.; Genova, F.; Jasniewicz, G.; Laloë, S.; Lesteven, S.; et al. The SIMBAD astronomical database. The CDS reference database for astronomical objects. *Astron. Astrophys. Suppl. Ser.* **2000**, *143*, 9–22. [CrossRef]

60. Barkhouse, W.A.; Hall, P.B. Quasars in the 2MASS Second Incremental Data Release. *Astron. J.* **2001**, *121*, 2843–2850. [CrossRef]

61. Shaw, M.S.; Romani, R.W.; Cotter, G.; Healey, S.E.; Michelson, P.F.; Readhead, A.C.S.; Richards, J.L.; Max-Moerbeck, W.; King, O.G.; Potter, W.J. Spectroscopy of Broad-line Blazars from 1LAC. *Astrophys. J.* **2012**, *748*, 49. [CrossRef]

62. Mao, L.S. 2MASS observation of BL Lac objects II. *New Astron.* **2011**, *16*, 503–529. [CrossRef]

63. Abazajian, K.N.; Adelman-McCarthy, J.K.; Agüeros, M.A.; Allam, S.S.; Allende Prieto, C.; An, D.; Anderson, K.S.J.; Anderson, S.F.; Annis, J.; Bahcall, N.A.; et al. The Seventh Data Release of the Sloan Digital Sky Survey. *Astrophys. J. Suppl. Ser.* **2009**, *182*, 543–558. [CrossRef]

64. Halpern, J.P.; Eracleous, M.; Mattox, J.R. Redshifts of Candidate Gamma-Ray Blazars. *Astron. J.* **2003**, *125*, 572–579. [CrossRef]

65. Enya, K.; Yoshii, Y.; Kobayashi, Y.; Minezaki, T.; Suganuma, M.; Tomita, H.; Peterson, B.A. JHK' Imaging Photometry of Seyfert 1 Active Galactic Nuclei and Quasars. I. Multiaperture Photometry. *Astrophys. J. Suppl. Ser.* **2002**, *141*, 23–29. [CrossRef]

66. Donato, D.; Ghisellini, G.; Tagliaferri, G.; Fossati, G. Hard X-ray properties of blazars. *Astron. Astrophys.* **2001**, *375*, 739–751. [CrossRef]

67. Jones, D.H.; Read, M.A.; Saunders, W.; Colless, M.; Jarrett, T.; Parker, Q.A.; Fairall, A.P.; Mauch, T.; Sadler, E.M.; Watson, F.G.; et al. The 6dF Galaxy Survey: Final redshift release (DR3) and southern large-scale structures. *Mon. Not. Royal Astron. Soc.* **2009**, *399*, 683–698. [CrossRef]

68. Kovalev, Y.Y.; Nizhelsky, N.A.; Kovalev, Y.A.; Berlin, A.B.; Zhekanis, G.V.; Mingaliev, M.G.; Bogdantsov, A.V. Survey of instantaneous 1–22 GHz spectra of 550 compact extragalactic objects with declinations from -30^{deg} to +43^{deg}. *Astron. Astrophy. Suppl.* **1999**, *139*, 545–554. [CrossRef]

 *symmetry*

Article

Lorentz-Violating Gravity Models and the Linearized Limit

Michael Seifert

Department of Physics, Astronomy, and Geophysics, Connecticut College, New London, CT 06320, USA;
mseifer1@conncoll.edu; Tel.: +1-860-439-5138

Received: 7 September 2018; Accepted: 10 October 2018; Published: 12 October 2018

Abstract: Many models in which Lorentz symmetry is spontaneously broken in a curved spacetime do so via a "Lorentz-violating" (LV) vector or tensor field, which dynamically takes on a vacuum expectation value and provides additional local geometric structure beyond the metric. The kinetic terms of such a field will not necessarily be decoupled from the kinetic terms of the metric, and will generically lead to a set of coupled equations for the perturbations of the metric and the LV field. In some models, however, the imposition of certain additional conditions can decouple these equations, yielding an "effective equation" for the metric perturbations alone. The resulting effective equation may depend on the metric in a gauge-invariant way, or it may be gauge-dependent. The only two known models yielding gauge-invariant effective equations involve differential forms; I show in this work that the obvious generalizations of these models do not yield gauge-invariant effective equations. Meanwhile, I show that a gauge-dependent effective equation may be obtained from any "tensor Klein–Gordon" model under similar assumptions. Finally, I discuss the implications of this work in the search for Lorentz-violating gravitational effects.

Keywords: Lorentz symmetry; linearized gravity; spontaneous symmetry breaking

---

1. Introduction

The prospect of Lorentz symmetry violation has received a fair amount of attention in recent years [1]. In the realm of particle physics on flat spacetime, the Standard Model Extension (SME) [2] has been developed as a descriptive framework for violations of Lorentz symmetry. The original Standard Model Lagrangian can be thought of as including all possible gauge-invariant and Lorentz-invariant combinations of local field operators up to some power-counting renormalizability cutoff, each with its own coefficient; it is then a matter of determining these coefficients via experiments. In general, a particular experiment will not be sensitive to a single coefficient in the Standard Model Lagrangian, but will instead place bounds on one or more combinations of these coefficients, with different experiments being sensitive to different combinations.

The SME "extends" the Standard Model by allowing the combinations of the field operators appearing in the Lagrange density to be Lorentz tensors instead of Lorentz scalars. The "minimal" SME contains all operators which are power-counting renormalizable; the "non-minimal" sector of the SME includes operators of higher powers as well. Since the overall Lagrange density must still be a scalar, this means that the "coefficients" of these new tensor operators must be tensors as well, contracted with the operators to form a Lorentz scalar. It then becomes an experimental question to measure the components of these coefficient tensors in some fiducial reference frame (usually taken to be the sun-centered reference frame [3].) As with the original Standard Model, a given experiment will be sensitive to some particular combinations of the components of the coefficient tensors. In the same sense that the original Standard Model action contains "all possible low-energy physics" that is consistent with the underlying gauge groups, locality, and Lorentz symmetry, the SME action contains "all possible low-energy physics" that is consistent with the underlying gauge groups and locality, but including frame-dependent effects. The general philosophy of the particle physics SME has been to remain agnostic concerning the underlying details of the "new physics" that might give rise to Lorentz-violating effects and simply focus on the phenomenological consequences that can arise from these violations; it is designed to be a framework rather than a model.

Given the utility of this framework in the arena of particle physics, one might wonder whether a similar framework could be devised for gravitational physics as well. After all, both particle physics and general relativity take Lorentz symmetry as one of their fundamental axioms, and a violation of Lorentz symmetry could show up in either realm. However, the path from the action to experimental signatures is significantly less straightforward in the gravitational sector than in the particle physics sector. As a result, it is not as clear that the existing framework for the "gravity sector of the SME" actually captures all possible signatures for Lorentz violation in the same way that the "particle physics sector of the SME" captures all possible signatures for Lorentz violation in particle physics.

The purpose of the present work is to argue that the range of models that can be analysed within the "minimal gravity sector of the SME", as it exists, is limited. In particular, there are only two known models in this category, and their obvious generalizations do not fall into the same category. In contrast, a broad class of models containing a Lorentz-violating field can be cast into a more general form which lies outside the minimal gravity sector of the SME. While it remains an open (and ill-defined) question whether these latter models are viable, these results suggest that a broader framework for Lorentz violation in the gravitational regime may be necessary.

2. Linearized Equations for Lorentz-Violating Gravity

A framework for the study of Lorentz symmetry violation in the context of gravity was developed by Kostelecký & Bailey [4], and much experimental work in the years since has searched for the effects described within this framework (see [5] for a review.) This framework attempts to follow the method taken by the particle physics sector of the SME by obtaining a Lorentz-violating equation of motion for the metric perturbations of a flat background, and then examining the phenomenology of the modified metrics. One begins by assuming a gravitational action of the form

$$ S = \int d^4x \sqrt{-g} \left[\frac{1}{2}(R + uR + s^{ab}R_{ab}^T + t^{abcd}C_{abcd}) + \mathcal{L}_{\text{mat}} \right]. \tag{1} $$

Here, the first term is the standard Einstein–Hilbert action (with $\Lambda = 0$); the second through fourth terms are "small" couplings of tensors u, s^{ab}, and t^{abcd} to the Ricci scalar, trace-free Ricci tensor, and Weyl tensor, respectively; and the final term encodes the dynamics of the fields other than the metric (we use reduced Planck units, i.e., $c = 8\pi G = \hbar = 1$). The tensors u, s^{ab}, and t^{abcd} are "coupled" to the spacetime curvature in this action, much as the particle physics SME coefficients are coupled to the fermion and boson fields. One would therefore expect that an action of this form could give rise to Lorentz-violating effects in gravitational physics. (As in [4], we restrict attention here to a Riemann spacetime with vanishing torsion, rather than the more general case of Riemann–Cartan spacetime. We also restrict attention to the "minimal" sector of the gravitational SME, in which higher derivatives of the Riemann tensor do not appear in the Lagrangian).

However, some important differences between this action and the particle physics SME action immediately arise. In particular, it is difficult to write down an action that includes a fixed background tensor ("prior geometry") in the context of a model with a dynamical metric. The equations of motion resulting from this prior geometry can easily violate the Bianchi identities, leading to a mathematically inconsistent model [6] (however, see [7] for a recent discussion of how these difficulties might be evaded).

For this reason, it has generally been assumed that the objects u, s^{ab}, and t^{abcd} in (1) are not fixed background tensors, but are instead constructed from some *dynamical* tensor which we denote as $\Psi^{\cdots}$ (the dots here denote a generic index structure). As this field is intended to dynamically provide us with the extra geometric background structure required for Lorentz symmetry violation, we refer to $\Psi^{\cdots}$ as the *Lorentz-violating field* (or *LV field*). Similar approaches have been proposed for the spontaneous breaking of another spacetime symmetry, namely scale-invariance [8,9]. However, as the fields used there were scalars, the issues outlined in this work do not arise.

In this picture, diffeomorphism symmetry is broken spontaneously: while the underlying action is assumed to be diffeomorphism-invariant, a certain class of solutions to the equations of motion have some or all of this symmetry broken. Since the SME focuses on local field operators, it is natural to assume that the underlying theory arises from a locally constructed Lagrangian as well. Finally, I assume that the full equations of motion resulting from this action are second-order. The condition that the action be locally constructed and diffeomorphism-invariant implies that the Lagrangian must be locally constructed from the metric, the Riemann tensor (and its derivatives), and the symmetrized derivatives of the other fields [10]; the condition that the equations of motion be second-order then prohibits any derivatives of the Riemann tensor or more than two derivatives of the other fields from appearing in any term in the Lagrangian. The full action is thus of the following form:

$$S = \int d^4x \sqrt{-g}\left[\frac{R}{2} + \frac{\xi}{2}\underbrace{f(\Psi^{\cdots})R^{\cdots}}_{\text{coupling}} + \frac{1}{2}\underbrace{(\nabla\Psi)(\nabla\Psi)}_{\text{kinetic}} + \underbrace{V(\Psi^{\cdots})}_{\text{potential}}\right]. \tag{2}$$

In (2), we divided the terms that could appear in a general action into three broad categories; the action (2) could, in principle, include multiple expressions of each type. The "kinetic" terms determine the dynamics of $\Psi^{\cdots}$; by choosing different combinations of cross-contractions and traces of the tensor $\nabla\Psi$, we can obtain different dynamics for the LV field. The "coupling" terms allow for a direct coupling between $\Psi^{\cdots}$ and the curvature, with each term multiplied by a coupling constant ξ_i. By adjusting the size of ξ, we can hope to "tune" the Lorentz-violating effects. Finally, the "potential" term consists of a potential energy $V(\Psi^{\cdots})$ for the field $\Psi^{\cdots}$. Each of these terms must be a spacetime scalar, with all of the indices in each term fully contracted.

To investigate the linearized limit of an action of the form (2), we must first vary the action with respect to the metric g_{ab} and the LV field $\Psi^{\cdots}$ to obtain the equations of motion. An important question arises: which of the three types of term (kinetic, coupling, or potential) contribute when we vary the action with respect to g_{ab}? It turns out that all three types of terms will vary when we vary the metric:

- The kinetic terms can vary in two ways when we change the metric. The more obvious way is that the full contraction of the tensor $\nabla\Psi$ with itself may require some raising and lowering of indices, which requires an implicit use of the metric or the inverse metric; these terms will change when the metric is varied. A more subtle point (but in some ways a more important one) is that the covariant derivative operator ∇_a also varies when the metric is varied. One could also consider Palatini-type theories, where the connection is viewed as a variable independent from the metric. As the correspondence between the "metric" and "Palatini" versions of a generalized gravity theory is not straightforward [11], I focus on the "metric" versions here for simplicity.
- The coupling terms explicitly depend on the Riemann tensor, which varies with the metric. As with the kinetic terms, they can also contain implicit factors of the metric or inverse metric that arise from index raising and lowering.
- The potential term must be a spacetime scalar constructed out of $\Psi^{\cdots}$; this will again usually require the raising and lowering of Ψ's indices, and so, implicitly depends on the metric.

The variation of the kinetic terms and coupling terms is particularly problematic. For the kinetic terms, the variation of the fields will yield something of the form (with indices temporarily suppressed)

$$\delta\left(\frac{1}{2}(\nabla\Psi)(\nabla\Psi)\right) \sim (\nabla\Psi)\left[\nabla(\delta\Psi) + \Psi(\nabla\delta g)\right] + (\nabla\Psi)(\nabla\Psi)(\delta g)$$

$$\sim -(\nabla\nabla\Psi)\delta\Psi + \left[-\Psi(\nabla\nabla\Psi) + (\nabla\Psi)(\nabla\Psi)\right]\delta g \tag{3}$$

which is integrated by parts in the second step. Note, in particular, that the coefficient of δg generically contains second derivatives of $\Psi^{\cdots}$; thus, second derivatives of $\Psi^{\cdots}$ will appear in the Einstein equation. Similarly, variation of the coupling terms will yield terms of the form

$$(\Psi R) \sim R(\delta\Psi) + \Psi(\delta R) \sim R(\delta\Psi) + \Psi(\nabla\nabla\delta g) \sim R(\delta\Psi) + (\nabla\nabla\Psi)\delta g. \tag{4}$$

All told, then, a generic action of the form (2) gives rise to equations of motion that are of schematically of the form

$$G + \xi F_1(\Psi)R + \Psi(\nabla\nabla\Psi)' + (\nabla\Psi)(\nabla\Psi) + F_2(\Psi) = 0, \tag{5a}$$

$$(\nabla\nabla\Psi) + \xi F_3(\Psi)R + F_4(\Psi) = 0, \tag{5b}$$

where G stands for the Einstein tensor, R stands for the Riemann tensor, and the F_i functions are some combinations (not necessarily invariants) of the LV field Ψ. The combinations of second derivatives that appear in the modified Einstein Equation (5a) will generically not be the same as those that appear in the equation of motion (5b) for Ψ itself; this is indicated by the use of the prime on the second derivatives of Ψ in the Einstein equation.

In the particle physics sector of the SME, the Lorentz-violating tensor coefficients are assumed to be constant throughout all of the flat spacetime, and to arise from the vacuum expectation value of the LV tensor field. We therefore require that there exists a solution to the equations of motion (5) with a flat metric ($g_{ab} = \eta_{ab}$) and a constant LV field ($\nabla\Psi = 0$). It is not too hard to see that if the metric is flat and the tensor field is constant, the equations of motion from (2) will reduce to the form

$$-\frac{1}{2}V(\Psi)\eta^{ab} + \left.\frac{\delta V}{\delta g_{ab}}\right|_{g\to\eta} = 0 \tag{6}$$

and

$$\left.\frac{\delta V}{\delta\Psi}\right|_{g\to\eta} = 0. \tag{7}$$

As a particularly simple example, suppose V is a single-argument function that depends only on a Lorentz invariant $X(\Psi)$ constructed from $\Psi^{\cdots}$ and the metric g_{ab}. In such an case, these equations become

$$-\frac{1}{2}V(X)\eta^{ab} + V'(X)\left.\frac{\delta X}{\delta g_{ab}}\right|_{g\to\eta} = 0 \tag{8}$$

and

$$V'(X)\left.\frac{\delta X}{\delta\Psi^{\cdots}}\right|_{g\to\eta} = 0, \tag{9}$$

and are satisfied if $V(X) = 0$ and $V'(X) = 0$. By constructing the potential appropriately, we can find models where there exist backgrounds consisting of a flat metric and a constant but non-zero LV field. We denote the background value of $\Psi^{\cdots}$ as $\bar{\Psi}^{\cdots}$.

The linearized limit of our model is found by looking at linearized perturbations about this background:

$$g_{ab} = \eta_{ab} + \epsilon h_{ab}, \qquad \Psi^{\cdots} = \bar{\Psi}^{\cdots} + \epsilon\tilde{\psi}^{\cdots}. \tag{10}$$

This results in a set of linearized equations of the form

$$(\delta G) + \zeta \bar{\Psi}(\delta R) + \bar{\Psi}\delta(\nabla\nabla\Psi)' + \ldots = 0, \tag{11a}$$

$$\delta(\nabla\nabla\Psi) + \zeta\delta R + \ldots = 0, \tag{11b}$$

where δG and δR stand for the linearized Einstein and Riemann tensors, respectively, and the ellipses in these equations contain terms that do not depend on the derivatives of h or $\tilde{\psi}$. Throughout this paper, we use δ to denote the linear-order perturbation of a quantity of the flat background.

Since our full Lagrangian (2) is diffeomorphism invariant, the linearized equations are invariant under the infinitesimal diffeomorphisms $h_{ab} \to h_{ab} + \mathcal{L}_v \eta_{ab}$ and $\tilde{\psi}^{\cdots} \to \tilde{\psi}^{\cdots} + \mathcal{L}_v\bar{\Psi}^{\cdots}$, where v_a is the vector field generating the diffeomorphism and $\mathcal{L}_v$ is the Lie derivative along this vector field. In terms of familiar index notation, these transformations are

$$h_{ab} \to h_{ab} + \partial_a v_b + \partial_b v_a \tag{12}$$

$$\tilde{\psi}^{a_1 \cdots a_k}{}_{b_1 \cdots b_l} \to \tilde{\psi}^{a_1 \cdots a_k}{}_{b_1 \cdots b_l} - \sum_{i=1}^{k} \bar{\Psi}^{a_1 \cdots c \cdots a_k}{}_{b_1 \cdots b_l}\partial_c v^{a_i} + \sum_{i=1}^{l} \bar{\Psi}^{a_1 \cdots a_k}{}_{b_1 \cdots c \cdots b_l}\partial_{b_i} v^c. \tag{13}$$

These can be thought of as gauge transformations for the linearized fields; perturbations of this form do not contain any physical information.

The ultimate goal in this process is to obtain a second-order partial differential equation for h alone, without reference to the LV fluctuations $\tilde{\psi}$. To do this, we must be able to eliminate the terms that are dependent on $\tilde{\psi}$ and its derivatives from (11a). We have certain tools at our disposal for this task. We may impose gauge conditions on h_{ab}, rewrite antisymmetrized derivatives of Ψ in terms of the Riemann tensor, or apply the linearized tensor equation (11b). We may also choose to restrict our solution space by imposing boundary conditions on various combinations of fields. Depending on the exact forms of the equations (11), we can envision three possibilities:

1. We can obtain a linearized second-order equation for h_{ab} that is manifestly gauge-invariant. Since the linearized Riemann tensor contains all of the "gauge-invariant information" encoded in the linearized metric, this means that the resulting equation is simply of the form

$$\delta G_{ab} + \zeta\bar{\Psi}^{\cdots}\delta R\ldots = 0. \tag{14}$$

    The process for obtaining this equation may (and, in the examples below, will) require boundary conditions to be set, but should not rely on any particular choice of gauge. Equations of this type are sometimes taken as the starting point for the study of Lorentz violation in gravitational physics [12,13], while remaining agnostic about the underlying model of Lorentz violation. However, it has also been pointed out that not all LV gravity models lead to such an equation [14].

2. We cannot obtain a linearized second-order equation for h_{ab} alone that is gauge-invariant, but we can find some particular gauge that allows us to eliminate $\tilde{\psi}$ from the Equation (11b). Again, the imposition of some appropriate set of boundary conditions may be necessary.

3. The Equations (11a) and (11b) are inextricably coupled. This means that we cannot write down a second-order equation for the metric alone without knowing the behavior of the LV field as well. In such a case, it could, in principle, still be possible to solve Equation (11b) for $\tilde{\psi}$ in terms of h and whatever other fields are present by Green's functions; one could then insert this solution for $\tilde{\psi}$ into (11a). However, the resulting equation would include some kind of integral over spacetime and would not result in a local second-order partial differential equation for h_{ab}.

As a shorthand, we can call these types of equations "Type I", "Type II", and "Type III", respectively. In general, it is not immediately clear from simple examination of the action (2) whether a particular model will yield equations of a particular type; one must generally work through the above process to find out. If we can obtain a Type I or Type II equation, we can, in principle, solve for the metric without having detailed knowledge of the LV field dynamics. In a model for which only Type III equations can be found, on the other hand, one must have knowledge of the LV field dynamics in order to figure out the weak-field limit; it is difficult to "remain agnostic" about the dynamics of the new fields (as one could with a Type I equation) and still make meaningful predictions.

3. *n*-Form Fields

As examples of models in which Type I equations can be obtained, I first consider the class of models where the LV field is an n-form field $\Psi_{a_1 \cdots a_n}$: a completely antisymmetric tensor of rank $(n, 0)$. For what follows, it is useful to denote a sequence of indices with braces, for example, we denote

$$\Psi^{a_1 \cdots a_n} \Psi_{a_1 \cdots a_n} \equiv \Psi^{\{a\}} \Psi_{\{a\}}. \tag{15}$$

The number of indices in a pair of braces should usually be evident from the context.

A particular feature of n-forms is that it is possible to define a natural notion of differentiation (the exterior derivative) that is independent of the metric. If $\Psi_{\{a\}}$ is an n-form, then $(d\Psi)_{\{a\}}$ is an $(n + 1)$-form defined in abstract-index notation as

$$(d\Psi)_{a\{b\}} = (n + 1)\nabla_{[a}\Psi_{\{b\}]}. \tag{16}$$

It is not hard to show that this quantity is independent of the derivative operator ∇_a used. One might imagine that this will make one's life easier; since the derivative is independent of the metric, the terms involving the second derivatives of the LV field in (5a)—and therefore in (11a)—will not arise. However, while this property is necessary for obtaining a decoupled equation for the metric, it is not sufficient.

We can use the exterior derivative to construct a kinetic term for a Lorentz-violating n-form field Ψ:

$$S = \int d^4x \sqrt{-g} \left[R - \frac{1}{2(n + 1)}(d\Psi)_{\{a\}}(d\Psi)^{\{a\}} - V(\Psi^2) + \mathcal{L}_{\text{coupling}} \right], \tag{17}$$

where $\Psi^2 \equiv \Psi_{\{a\}}\Psi^{\{a\}}$ and

$$\mathcal{L}_{\text{coupling}} = \xi_1 \Psi^{ab}{}_{\{e\}}\Psi^{cd\{e\}} R_{abcd} + \xi_2 \Psi^a{}_{\{c\}}\Psi^{b\{c\}} R_{ab} + \xi_3 \Psi^2 R. \tag{18}$$

In what follows, we also find it necessary to require that

$$4\xi_1 + (n - 1)\xi_2 = 0. \tag{19}$$

Both the bumblebee model described in [6] and the tensor models described in [15] fall into this category, using a one-form or a two-form (respectively) as the LV field. (The models in [15] can also include a potential that depends on the invariant $\Psi_{ab}\Psi_{cd}\epsilon^{abcd}$ as well as the tensor norm $\Psi_{ab}\Psi^{ab}$. In general, such an invariant only exists if $n = d/2$, where d is the dimension of the spacetime. The arguments which follow can be generalized to such cases straightforwardly). In the former work, the ξ_1 term does not exist and ξ_2 is unconstrained; in the latter work, the condition (19) (with $n = 2$) is found to be necessary to obtain a "decoupled" effective equation. The generalized bumblebee models described in [14] are also of this general form, with the modification that the indices in the kinetic term for the one-form Ψ_a are contracted using an "effective metric" constructed out of the spacetime metric g_{ab} and Ψ_a itself.

To understand how the full decoupling of the equations can be accomplished in this case, we must first look at the linearized equation of motion for the LV field. This is

$$\partial_d(d\bar{\psi})^{dab\{c\}} - \lambda\Psi^{ab\{c\}}\delta(\Psi^2) + 2\xi_1 \Psi^{de[\{c\}}\delta R^{ab]}{}_{de} + 2\xi_2 \Psi^{d[b\{c\}}\delta R^{a]}{}_d + 2\xi_3 \Psi^{ab\{c\}}\delta R = 0, \tag{20}$$

where $\lambda \equiv V''(\Psi^2)$, $\delta(\Psi^2)$ is the linear variation in Ψ^2, and the tensor $(d\psi)_{\{a\}} \equiv \delta(d\Psi)_{\{a\}}$ is defined similarly to (16). (Note that for any tensor $t_{\{b\}}$ which vanishes in the background, $\nabla_a t_{\{b\}} = \partial_a t_{\{b\}} + \mathcal{O}(\epsilon^2)$).

In general, the derivatives present in the linearized LV equation of motion (11b) (Equation (20) in the present case) are the only tools we have to eliminate the derivatives of $\tilde{\psi}$ in the linearized Einstein Equation (11a). However, the structure of the kinetic term for an n-form field leads to two auxiliary conditions which turn out to be crucial for the decoupling. First, if we take the divergence of (20), the kinetic term identically vanishes and we obtain

$$- \lambda \Psi^{ab\{c\}} \partial_a \delta(\Psi^2) + 2\xi_1 \Psi^{de[\{c\}} \partial_a \delta R^{ab]}{}_{de} + 2\xi_2 \Psi^{d[b\{c\}} \partial_a \delta R^{a]}{}_d + 2\xi_3 \Psi^{ab\{c\}} \partial_a \delta R = 0. \tag{21}$$

However, the Riemann tensor satisfies the Bianchi identity $\nabla_{[a} R_{bc]de} = 0$. At linear order, this becomes $\partial_{[a} \delta R_{bc]de} = 0$, and taking the traces of this equation implies that $2\partial_{[a}\delta R_{b]c} = \partial^d \delta R_{dcab}$ and $2\partial^a \delta R_{ab} = \partial_b \delta R$. Splitting up the antisymmetrizers in (21) and applying the Bianchi identities lets us rewrite the second and third terms as

$$2\xi_1 \Psi^{de[\{c\}} \partial_a \delta R^{ab]}{}_{de} + 2\xi_2 \Psi^{d[b\{c\}} \partial_a \delta R^{a]}{}_d$$
$$= \frac{\xi_2}{n} \Psi^{dbc_1 \cdots c_{n-2}} \partial_d \delta R + \frac{2[4\xi_1 + (n-1)\xi_2]}{n(n-1)} \left[\Psi^{adc_1 \cdots c_{n-2}} \partial_{[a} \delta R_{d]}{}^b - \sum_{i=1}^{n-2} \Psi^{adc_1 \cdots b \cdots c_{n-2}} \partial_a \delta R_d{}^{c_i} \right]. \tag{22}$$

Since we assume the condition (19) applies to the coupling constants, this means that (21) simplifies greatly:

$$\Psi^{ab\{c\}} \partial_a \left[-\lambda \delta(\Psi^2) + \left(\frac{\xi_2}{n} + 2\xi_3 \right) \delta R \right] = 0. \tag{23}$$

A possible solution to this equation (though not the only one) is that the quantity in square brackets vanishes:

$$\lambda \delta(\Psi^2) = \left(\frac{\xi_2}{n} + 2\xi_3 \right) \delta R. \tag{24}$$

I call this the "massive mode condition", since it effectively requires that $\delta(\Psi^2)$ only deviates by a small amount (i.e., $\mathcal{O}(\xi_i)$) from zero. Note, however, that this condition can only be applied if (19) holds; for a general set of coupling constants, imposing such a condition would be more poorly motivated.

The second condition can be found by taking the "curl" of (20). Since we now know that we can choose a solution such that $\lambda \delta(\Psi^2)$ is $\mathcal{O}(\xi_i)$, the linearized Equation (20) reduces to

$$\partial^b (d\tilde{\psi})_{b\{a\}} = \mathcal{O}(\xi_i). \tag{25}$$

If we take the antisymmetrized derivative of the left-hand side, we obtain

$$\partial_{[c} \partial^b (d\tilde{\psi})_{|b|a_1 \cdots a_n]} = \partial_{[c} \partial^b \partial_{|b|} \tilde{\psi}_{a_1 \cdots a_n]} - n\partial_{[c} \partial^b \partial_{a_1} \tilde{\psi}_{|b|a_2 \cdots a_n]} = \Box \partial_{[c} \tilde{\psi}_{a_1 \cdots a_n]}, \tag{26}$$

where the second term in the middle step vanishes due to the antisymmetrization over the partial derivatives. Thus, we have

$$\Box (d\tilde{\psi}_{\{a\}}) = \mathcal{O}(\xi_i). \tag{27}$$

As an aside, the identity (26) can also be derived elegantly using the language of differential forms. Define $\underline{d}$ and $\underline{\delta}$ as the exterior derivative and codifferential on the manifold (note that this $\underline{\delta}$ is an operator that takes n-forms to $(n-1)$-forms, and is distinct from the usage of δ to denote linearized perturbations in this work.) The kinetic term in (20) can be written as $\underline{\delta}\underline{d}\psi$. We then have $\underline{d}\underline{\delta}(\underline{d}\psi) = (\underline{d}\underline{\delta} + \underline{\delta}\underline{d})(\underline{d}\psi)$, since $\underline{d}^2 = 0$. This operator acting on $(\underline{d}\psi)$ is the so-called *Laplace–de Rham operator*, which differs from the "conventional" tensor Laplacian $\Box = \nabla^a \nabla_a$ by (at most) various contractions of the curvature with the n-form.

Equation (27) is a hyperbolic differential equation on our manifold, and via an appropriate choice of boundary conditions, we can require that the linearized field strength $(d\tilde{\psi})_{\{a\}}$ remains $\mathcal{O}(\xi_i)$ throughout our spacetime:

$$d\tilde{\psi}_{\{a\}} = \mathcal{O}(\xi_i). \tag{28}$$

I call this condition the "curl condition". As with the massive mode condition (24), it must be noted that this condition is not true for the most general solution of the equation (27) from which it stems.

With these in mind, we can now attempt to decouple the n-form fluctuations from the metric fluctuations. The linearized Einstein equations in this case are of the form

$$\delta G_{ab} - n\lambda\delta(\Psi^2)\Psi_a{}^{\{c\}}\Psi_{b\{c\}} + \sum_i \xi_i \left[(\mathcal{A}_{(i)})_{ab}{}^{cdef}\delta R_{cdef} + (\mathcal{B}_{(i)})_{ab}{}^{cdef\{g\}}\delta(\nabla_{(c}\nabla_{d)}\Psi_{ef\{g\}}) \right] = 0, \tag{29}$$

where the $(\mathcal{A}_{(i)})_{ab}{}^{cdef}$ and $(\mathcal{B}_{(i)})_{ab}{}^{cdef\{g\}}$ tensors are constructed out of the background fields $\Psi_{\{a\}}$ and η_{ab}. (Recall from above that the antisymmetrized second derivative of $\Psi_{\{a\}}$ can be written in terms of contractions of the Riemann tensor with $\Psi_{\{a\}}$.) It is important here to note that there are no terms that arise involving the second derivatives of $\Psi_{\{a\}}$ at $\mathcal{O}(\xi^0)$. For a generic model, such terms would arise from the variation in the kinetic term for the LV field, since the covariant derivative operator is dependent on the metric. However, the field strength (16) is independent of the metric, and so it evades this obstacle.

Assuming the massive mode condition (24) holds, we can replace the second term in (29) with curvature terms multiplied by coefficients of $\mathcal{O}(\xi_i)$, and the third term is already of the desired form. It is the second derivatives of $\Psi_{\{a\}}$ that arise in (29) (the terms contracted with $(\mathcal{B}_{(i)})_{ab}{}^{cdef\{g\}}$) that we need to eliminate. Explicitly, these tensors are

$$(\mathcal{B}_{(1)})_{ab}{}^{cdef\{g\}} = 4\Psi_{(a}{}^{(c|\{g\}}\eta_{b)}{}^e\eta^{|d)f} \tag{30}$$

$$(\mathcal{B}_{(2)})_{ab}{}^{cdef\{g\}} = \eta_{ab}\Psi^{(c|f\{g\}}\eta^{|d)e} - \Psi^{cf\{g\}}\eta^d{}_{(a}\eta^e{}_{b)} + \Psi_{(a}{}^{f\{g\}}\eta^{cd}\eta_{b)}{}^e - \Psi_{(a}{}^{f\{g\}}\eta^{ce}\eta_{b)}{}^d, \tag{31}$$

$$(\mathcal{B}_{(3)})_{ab}{}^{cdef\{g\}} = 2\Psi^{ef\{g\}}(\eta_{ab}\eta^{cd} - \eta_{(a}{}^c\eta_{b)}{}^d). \tag{32}$$

If we can show that these tensors contracted with $\delta(\nabla_{(c}\nabla_{d)}\Psi_{ef\{g\}})$ are automatically $\mathcal{O}(\xi_i)$, then we will be able to write down an approximate equation, correct to $\mathcal{O}(\xi_i)$, that only contains various contractions of the linearized Riemann tensor with the background values of $\Psi_{\{a\}}$. This is the desired form of the equations of motion underlying the gravitational sector of the SME.

Enforcing the massive mode condition (24), under which $\delta\Psi^2$ is $\mathcal{O}(\xi_i)$, implies that

$$\frac{1}{2}\delta(\nabla_a\Psi^2) = \delta(\Psi^{\{b\}}\nabla_a\Psi_{\{b\}}) = \Psi^{\{b\}}\delta(\nabla_a\Psi_{\{b\}}) \tag{33}$$

is $\mathcal{O}(\xi_i)$ as well. This allows us to eliminate any terms where we have contracted $\Psi^{ef\{g\}}$ with $\delta(\nabla_{(c}\nabla_{d)}\Psi_{ef\{g\}})$ in the linearized equations of motion. In particular, this guarantees that both terms arising from (32) do not contribute to the equations of motion at $\mathcal{O}(\xi_i)$.

Enforcing the curl condition (28) allows us to eliminate more of the above derivatives. We have

$$\Psi^{b_1\cdots b_n}(d\tilde{\psi})_{ab_1\cdots b_n} = \frac{1}{n+1}\Psi^{b_1\cdots b_n}\left[\delta(\nabla_a\Psi_{b_1\cdots b_n}) - n\delta(\nabla_{[b_1}\Psi_{|a|b_2\cdots b_n]})\right]. \tag{34}$$

Since we are assuming that $(d\tilde{\psi})_{\{a\}}$ is $\mathcal{O}(\xi_i)$ throughout the spacetime, and since the first term on the right-hand side is also $\mathcal{O}(\xi_i)$ via the massive mode condition, we conclude that

$$\Psi^{b_1\cdots b_n}\delta(\nabla_{[b_1}\Psi_{|a|b_2\cdots b_n]}) = \mathcal{O}(\xi_i) \tag{35}$$

as well. We conclude that any contraction of the form $\Psi^{cf\{g\}}$ or $\Psi^{df\{g\}}$ (and by symmetry, $\Psi^{ce\{g\}}$ or $\Psi^{de\{g\}}$) with $\delta(\nabla_{(c}\nabla_{d)}\Psi_{ef\{g\}})$ will also be of the form $\mathcal{O}(\xi_i)$; this eliminates the first two terms of (31).

What remains are (30) and the last two terms of (31), and the only tool we have not yet used is the equation of motion itself. The kinetic term in (20) can be rewritten as

$$\partial^p(d\tilde{\psi})_{pqr\{s\}} = (n+1)\delta(\nabla_c\nabla_d\Psi_{ef\{g\}})\eta^{cp}\eta_{[p}{}^d\eta_q{}^e\eta_r{}^f\eta_{s_1\cdots s_{n-2}]}^{g_1\cdots g_{n-2}}, \tag{36}$$

where $\eta_{s_1\cdots s_n}^{g_1\cdots g_n} \equiv \eta_{s_1}{}^{g_1}\cdots\eta_{s_n}{}^{g_n}$. If we contract (36) with the tensor $\eta_{(a}^q\Psi_{b)}{}^{r\{s\}}$, then we obtain the tensor $\delta(\nabla_c\nabla_d\Psi_{ef\{g\}})$ contracted with the tensor

$$(n+1)\left[\eta^{cp}\eta_{(a}{}^q\Psi_{b)}{}^{r\{s\}}\right]\eta_{[p}{}^d\eta_q{}^e\eta_r{}^f\eta_{\{s\}]}{}^{\{g\}}$$

$$= \frac{1}{n!}\left(2\eta^{cd}\eta_{(a}{}^{[e}\Psi_{b)}{}^{f]\{g\}} - 2\eta^{c[e}\eta_{(a}{}^{|d|}\Psi_{b)}{}^{f]\{g\}} + 2\eta^{c[e}\eta_{(a}{}^{f]}\Psi_{b)}{}^{d\{g\}} + \ldots\right), \tag{37}$$

where the ellipses stand for terms involving permutations where two or more of the indices d, e, and f are attached to the background LV field Ψ, along with one of the free indices a or b. Such terms only exist if $n > 2$. These latter terms do not arise in linearized Einstein Equations (29). Further, they cannot be eliminated via the auxiliary conditions (33) or (35), as these conditions require Ψ to be fully contracted with the perturbation $\delta(\nabla_c\nabla_d\Psi_{ef\{g\}})$. Thus, a Lorentz-violating model including an n-form LV field with $n > 2$ cannot be analysed in the SME gravitational framework.

The exceptions are the lowest-rank cases of $n = 1$ and $n = 2$. For $n = 1$, the ξ_1 term (30) does not exist, and the kinetic term in the LV equation of motion contracted with $\eta_{(a}{}^c\Psi_{b)}$ is simply

$$(\eta^{cd}\eta_{(a}{}^e\Psi_{b)} - \eta^{ce}\eta_{(a}{}^d\Psi_{b)})\delta(\nabla_c\nabla_d\Psi_e). \tag{38}$$

These terms are exactly the remaining terms in (31) in this case, and so all non-curvature terms in (29) are $\mathcal{O}(\xi^2)$ or higher. For $n = 2$, the remaining terms in (30) and (31) can be combined to equal the terms explicitly written out in (37) (recalling from (19) that $4\xi_1 = \xi_2$ in this case.) The remaining terms in (20) are all of the form $\mathcal{O}(\xi_i)$, either explicitly or via the massive mode condition, and thus, all non-curvature terms in (29) are $\mathcal{O}(\xi_i^2)$ or higher.

It is worth taking stock of all of the assumptions that have gone into obtaining these effective equations. Only in two cases ($n = 1$, and $n = 2$ in the case $4\xi_1 = \xi_2$) can this procedure be completed. In both cases, auxiliary conditions concerning the solutions—namely, (24) and (28)—are required to obtain a linearized metric equation that fully eliminates the LV field perturbations from the linearized Einstein equation. A general linearized solution will not satisfy these conditions, and will therefore not obey the effective equation derived by this procedure. Moreover, while the arguments leading to these auxiliary conditions can be generalized to higher-rank n-forms, the conditions are insufficient to eliminate the LV field perturbations from the equations of motion. It seems unlikely that a general action for a LV n-form field (17) can be reduced to an effective equation of the form (14).

4. Tensor Klein–Gordon Fields

It may occur that we cannot write an analog of the linearized Einstein equation solely in terms of the linearized Riemann tensor in a model containing an LV field. However, we can still hope to find an equation of motion for the linearized metric that, while not invariant under the "gauge transformations" $h_{ab} \rightarrow h_{ab} + \partial_{(a}\xi_{b)}$, is at least decoupled from any explicit dependence on the fluctuations of the LV field. Such an equation is what I called a "Type II" equation above.

An example of a class of models that can yield such equations are the "tensor Klein–Gordon" actions, of the form

$$S = \int d^4x\sqrt{-g}\left[R - \frac{1}{2}\nabla_a\Psi_{\{b\}}\nabla^a\Psi^{\{b\}} - V(\Psi^2)\right], \tag{39}$$

where $\Psi_{\{b\}}$ is now an arbitrary rank-$(n,0)$ tensor field without any particular symmetry structure. Note the absence of any particular explicit coupling between the LV field $\Psi_{\{b\}}$ and the Riemann tensor in the Lagrangian. The linearized equation of motion for the LV field in such a model is

$$\eta^{cd}\delta(\nabla_c\nabla_d\Psi_{\{a\}}) - 2\lambda\Psi_{\{a\}}\delta(\Psi^2) = 0, \tag{40}$$

with $\lambda = V''(\bar{\Psi}^2)$, as before. In terms of the perturbations $\tilde{\psi}_{\{a\}}$ and h_{ab}, the first term above becomes

$$\eta^{cd}\delta(\nabla_c\nabla_d\Psi_{\{a\}}) = \eta^{cd}\left(\partial_c\partial_d\tilde{\psi}_{\{a\}} - \sum_i \partial_c\Gamma^f{}_{da_i}\bar{\Psi}_{a_1...b...a_n}\right)$$
$$= \Box\left(\tilde{\psi}_{\{a\}} - \frac{1}{2}\sum_i \bar{\Psi}_{a_1...}{}^b{}_{...a_n}h_{a_ib}\right) - \sum_i \partial^c\partial_{[a_i}h_{b]c}\bar{\Psi}_{a_1...}{}^b{}_{...a_n}. \tag{41}$$

However, in the standard harmonic gauge $\partial^b h_{ab} = \frac{1}{2}\partial_a h$ (where $h \equiv h^a{}_a$), we have

$$\partial^c\partial_{[a_i}h_{b]c} = \partial_{[a_i}\partial^c h_{b]c} = \frac{1}{2}\partial_{[a_i}\partial_{b]}h = 0. \tag{42}$$

Moreover, in terms of the perturbations, we have

$$\delta(\Psi^2) = 2\bar{\Psi}^{\{a\}}\tilde{\psi}_{\{a\}} - \sum_i \bar{\Psi}^{a_1...a_i...a_n}\bar{\Psi}_{a_1...}{}^b{}_{...a_n}h_{a_ib} = 2\bar{\Psi}^{\{a\}}\left(\tilde{\psi}_{\{a\}} - \frac{1}{2}\sum_i \bar{\Psi}_{a_1...}{}^b{}_{...a_n}h_{a_ib}\right). \tag{43}$$

(The negative sign arises from $\delta(g^{ab}) = -\eta^{ac}\eta^{bd}h_{cd}$.) Putting this all together, the linearized equation of motion for the LV field is (in the harmonic gauge)

$$\Box\left(\tilde{\psi}_{\{a\}} - \frac{1}{2}\sum_i \bar{\Psi}_{a_1...}{}^c{}_{...a_n}h_{a_ic}\right) - 4\lambda\bar{\Psi}_{\{a\}}\bar{\Psi}^{\{b\}}\left(\tilde{\psi}_{\{b\}} - \frac{1}{2}\sum_i \bar{\Psi}_{b_1...}{}^c{}_{...b_n}h_{b_ic}\right) = 0. \tag{44}$$

This is a hyperbolic equation for the quantity in parentheses above, which means that if we choose the appropriate boundary conditions, we can make the said quantity as small as we want:

$$\tilde{\psi}_{\{a\}} - \frac{1}{2}\sum_i \bar{\Psi}_{a_1...}{}^c{}_{...a_n}h_{a_ic} \approx 0. \tag{45}$$

In other words, via a choice of boundary condition, the LV field perturbations $\tilde{\psi}_{\{a\}}$ can be written in terms of the background tensor values $\bar{\Psi}_{\{a\}}$ and the metric perturbations h_{ab}. The LV field perturbations can therefore be completely eliminated from the linearized Einstein Equation (11a) in favor of the metric perturbations. The result will be an equation which holds only in harmonic gauge, but which does not contain any explicit dependence on the LV field fluctuations. While such an equation will not fit into the gravitational framework of the SME, one could still analyze the resulting equation in terms of its predicted post-Newtonian effects, its predictions for gravitational wave propagation, and so forth. The linearized metric found in Section IV.B of [4], for example, is the linearized metric for a model containing a Lorentz-violating "vector Klein–Gordon" field, and in obtaining that metric, the same assumption (45) concerning the perturbations of the LV field is made.

This result for tensor Klein–Gordon models could be generalized to other gauges; for example, the result (42) would still hold in which $\partial^b\bar{h}_{ab}$ (where $\bar{h}_{ab} = h_{ab} - \frac{1}{2}\eta_{ab}h$) is equal to the gradient of some scalar function f, rather than vanishing as assumed in the harmonic gauge. Imposing a different gauge choice would yield an effective equation of motion for h_{ab} that appears superficially different from the equation obtained in the original gauge. However, the solutions to this new equation should simply be the solutions of the original equation modified by a gauge transformation; the choice of gauge would not cause a change in the physically measurable aspects of the metric.

Finally, we note that as with the differential forms in Section 3, the most general solutions to the linearized equations of motion will not obey (45), and will therefore not necessarily obey the effective equation derived from it.

5. Discussion

In general, the Euler–Lagrange equations for a model containing both a tensor field and a dynamical metric contain coupled kinetic terms, and if the tensor field takes on a vacuum expectation value in flat spacetime, this coupling will persist in the linearized equations about this background. One can try to evade this by using an n-form field for which the "natural" kinetic term for the field (involving the tensor norm of the field strength) does not couple directly to the metric perturbations. With such an equation of motion for the linearized metric in hand, one could proceed to follow the analysis of [4], looking for the observational signatures of Lorentz violation of the types derived there. However, this assumption is not sufficient to fully "decouple" the equations; one must also restrict attention to the solutions satisfying (24) and (35), and even in such instances, the decoupling only occurs when $n = 1$ or $n = 2$.

Even granting these assumptions, it appears that the great majority of models one can write down in which a vector or tensor field takes on a vacuum expectation value do not yield Type I equations. To date, only the two above-mentioned models have been found to yield equations of Type I, and their obvious generalizations to higher-rank differential forms fail to yield a decoupled, gauge-invariant equation for the linearized metric. (One could generalize the two-form fields in [15] in the same way that the "generalized bumblebee models" of [14] generalized the original bumblebee model [6]; see the discussion following (18). It seems likely that such a model could yield Type I equations as well, though this remains to be shown rigorously.) If the linearized equations are Type II or III, the form of the linearized metric will almost certainly differ from the predicted linearized metric in the SME, and thus, the experimental signatures of this model will not match those of the SME. Such models would have to be analyzed on a case-by-case basis from first principles. Indeed, in [4], the second derivatives of the fields that cannot be eliminated are implicitly placed into a divergence-free tensor Σ_{ab}. It is conjectured in that work that models in which Σ_{ab} does not vanish "may even be generic"; the current work lends credence to this conjecture.

In the particle physics sector of the SME, the size of the coefficients of the Lorentz-violating terms in the action bears directly on the size of the effects that are physically observed. One would therefore naïvely expect that an action of the form (2) containing a "small" coupling term would lead to "small" violations of Lorentz symmetry. For the known models [6,14,15] that yield Type I equations, this is effectively the case—small values of the coefficients ξ_i lead to small contributions to the linearized equations of motion.

In contrast, the size of the Lorentz-violating effects in a model yielding a Type II or Type III equation may not be determined by the size of the "coupling" terms in the action. For example, the tensor Klein–Gordon models described in Section 4 have no coupling terms at all, but give rise to an effective equation with a kinetic term of the form

$$\Box h_{ab} + \overset{\cdots}{\Psi}\,\overset{\cdots}{\Psi}\,\mathcal{D}_{\ldots}[h] + \cdots = 0, \tag{46}$$

where $\mathcal{D}_{\ldots}[h]$ is some second-order differential operator acting on h_{ab}, and the indices of the two $\overset{\cdots}{\Psi}$ tensors are contracted with this operator in some way. The difference between this equation of motion and the linearized Einstein equation is therefore governed by coefficients of $\mathcal{O}(E^2)$, where E is the energy scale of the LV field's vacuum expectation value in reduced Planck units. Since these deviations would increase with the magnitude of the background field, one could, in principle, completely exclude such models via both precision gravitational experiments (thereby excluding sufficiently large energy scales) and accelerator experiments (excluding sufficiently small energy scales).

Of course, the gap between "sufficiently large" and "sufficiently small" may be quite large indeed. As a very rough estimate, one might assume that a linearized metric solving (46) would differ from the weak-field GR solution by deviations of $\mathcal{O}(E^2)$. Given that spacetime-anisotropic deviations from the weak-field GR solution are currently bounded to about one part in 10^{11} [1], this implies that the vacuum expectation value for Ψ could be no larger than about $10^{-5} m_P \approx 10^{12}$ GeV, nowhere near the current bounds from accelerator experiments.

It is also possible that a model that yields Type II or Type III equations will be phenomenologically non-viable for reasons other than the form of its linearized metric. The tensor Klein–Gordon fields in Section 4, for example, include a kinetic term that is of indeterminate sign, and do not contain any gauge degrees of freedom. This is likely to lead to a Hamiltonian that is unbounded below, running a strong risk of runaway solutions. Thus, while it may be possible to find a linearized metric in a Type II model (as was done for the vector Klein–Gordon model in [4]), such a solution would need to be checked for stability. Similarly, while a Type II equation can be reduced to a single equation involving the metric perturbation h_{ab}, the number of physical degrees of freedom of the metric may be different from those of a metric obeying a Type I equation; the lack of gauge symmetry in a Type II equation means that degrees of freedom that correspond to gauge transformations in a Type I equation may now become physically meaningful. The extra propagating modes arising in such a model could be used to constrain it or rule it out.

One might conjecture that all viable Lorentz-violating models of gravity yield equations of Type I. Unless we simply *define* a "viable model" to be one that can yield linearized equations of Type I—which is somewhat tautological, and requires clear motivation—we must have a clear idea of what we mean by "viable". The criteria for a model to be viable on purely theoretical grounds are something of a judgement call. Most physicists would probably agree that a model with a Hamiltonian that is unbounded below is probably not a good one, but whether or not one should be interested in a model with extra metric degrees of freedom is more open to debate. Many other possible criteria for a "viable" model could be proposed, depending on the context one is interested in. Moreover, such a conjecture elides the fact that a given model may have some linearized solutions that obey such an equation and some that do not; in the known examples [6] and [15], extra conditions on the linearized solutions are needed to obtain a "Type I" equation. Without a more precise and well-motivated statement of the properties of a viable model, and justifications for any necessary auxiliary conditions on the linearized solutions, this "conjecture" is really more of a hope.

The SME framework has been incredibly useful for discussing the signatures of Lorentz violation in the realm of particle physics. Any self-consistent model of Lorentz violation in particle physics can be mapped onto one or more coefficients in the SME action (either in the minimal or the non-minimal versions of the framework), and there is a well-developed machinery that maps these coefficients into experimental results. In the realm of particle physics, therefore, the SME can legitimately claim to be a mature framework for the study of violations of Lorentz symmetry.

In contrast, the gravitational sector of the SME does not yet have this same status. Under certain assumptions, certain n-form models mentioned above do fit within the SME gravity framework developed in [4]. However, if one starts with a particular action involving a dynamical metric and a tensor field with a vacuum expectation value, one cannot easily discern which of the above three types it falls into. There is no way to determine this other than to go through the long process of deriving the perturbational equations of motion (11) and trying to eliminate the LV tensor fluctuations from the linearized Einstein equation. Even then, the process is not guaranteed to be straightforward. It may rely on subtle auxiliary conditions (such as the massive mode condition (24) or the curl condition (28)) that are not immediately evident from the linearized equations of motion and that a general linearized solution of the equations of motion may not satisfy. Finally, there is no guarantee that any particular model will result in a linearized equation that yields the same linearized gravity phenomenology as a Type I equation; the experimental signatures of Type II and Type III models can, in principle, be totally different from those of the well-studied Type I models. It is therefore unclear that the gravitational sector of the SME has the same level of phenomenological universality that the particle physics sector of the SME enjoys.

Funding: This research received no external funding.

Acknowledgments: The author would like to acknowledge the helpful and thorough comments of one of the anonymous reviewers.

References

1. Kostelecký, V.A.; Russell, N. Data tables for Lorentz and CPT violation. *Rev. Mod. Phys.* **2011**, *83*, 11–31. [CrossRef]
2. Colladay, D.; Kostelecký, V.A. Lorentz-violating extension of the standard model. *Phys. Rev. D* **1998**, *58*, 116002. [CrossRef]
3. Bluhm, R.; Kostelecký, V.A.; Lane, C.D.; Russell, N. Probing Lorentz and CPT violation with space-based experiments. *Phys. Rev. D* **2003**, *68*, 125008. [CrossRef]
4. Bailey, Q.; Kostelecký, V. Signals for Lorentz violation in post-Newtonian gravity. *Phys. Rev. D* **2006**, *74*, 045001. [CrossRef]
5. Tasson, J.D. The standard-model extension and gravitational tests. *Symmetry* **2016**, *8*, 111. [CrossRef]
6. Kostelecký, V.A. Gravity, Lorentz violation, and the standard model. *Phys. Rev. D* **2004**, *69*, 105009. [CrossRef]
7. Bluhm, R. Explicit versus spontaneous diffeomorphism breaking in gravity. *Phys. Rev. D* **2015**, *91*, 065034. [CrossRef]
8. Guendelman, E.I. Scale invariance, new Inflation and decaying Λ-terms. *Mod. Phys. Lett. A* **1999**, *14*, 1043–1052. [CrossRef]
9. Guendelman, E.I.; Nishino, H.; Rajpoot, S. Two-measure approach to breaking scale-invariance in a standard-model extension. *Phys. Lett. Sect. B Nuclear Elem. Part. High-Energy Phys.* **2017**, *765*, 251–255. [CrossRef]
10. Iyer, V.; Wald, R.M. Some properties of the Noether charge and a proposal for dynamical black hole entropy. *Phys. Rev. D* **1994**, *50*, 846–864. [CrossRef]
11. Iglesias, A.; Kaloper, N.; Padilla, A.; Park, M. How (not) to use the Palatini formulation of scalar-tensor gravity. *Phys. Rev. D Part. Fields Gravit. Cosmol.* **2007**, *76*, 1–11. [CrossRef]
12. Kostelecký, V.A.; Tasson, J.D. Constraints on Lorentz violation from gravitational Čerenkov radiation. *Phys. Lett. Sect. B Nucl. Elem. Part. High-Energy Phys.* **2015**, *749*, 551–559. [CrossRef]
13. Kostelecky, V.A.; Mewes, M. Testing local Lorentz invariance with gravitational waves. *Phys. Lett. Sect. B Nuclear Elem. Part. High-Energy Phys.* **2016**, *757*, 510–514. [CrossRef]
14. Seifert, M.D. Vector models of gravitational Lorentz symmetry breaking. *Phys. Rev. D* **2009**, *79*, 124012. [CrossRef]
15. Altschul, B.; Bailey, Q.G.; Kostelecký, V.A. Lorentz violation with an antisymmetric tensor. *Phys. Rev. D* **2010**, *81*, 065028. [CrossRef]

Article

Relating Noncommutative SO(2,3)$_\star$ Gravity to the Lorentz-Violating Standard-Model Extension

Quentin G. Bailey [1] and Charles D. Lane [2,3,*]

[1]   Physics Department, Embry-Riddle Aeronautical University, Prescott, AZ 86301, USA; baileyq@erau.edu
[2]   Department of Physics, Berry College, Mount Berry, GA 30149, USA
[3]   IU Center for Spacetime Symmetries, Indiana University, Bloomington, IN 47405, USA
[*]   Correspondence: clane@berry.edu; Tel.: +1-706-290-2673

Received: 3 September 2018; Accepted: 3 October 2018; Published: 11 October 2018

Abstract: We consider a model of noncommutative gravity that is based on a spacetime with broken local SO(2,3)$_\star$ symmetry. We show that the torsion-free version of this model is contained within the framework of the Lorentz-violating Standard-Model Extension (SME). We analyze in detail the relation between the torsion-free, quadratic limits of the broken SO(2,3)$_\star$ model and the Standard-Model Extension. As part of the analysis, we construct the relevant geometric quantities to quadratic order in the metric perturbation around a flat background.

Keywords: Lorentz violation; noncommutative geometry; gravity

---

1. Introduction

While noncommutative geometry has been studied for more than 70 years [1], it has been especially popular as a possible framework for physics beyond the Standard Model in recent decades [2,3]. In particular, several extensions to general relativity that incorporate noncommutative geometry have been proposed [4–10]. In this paper, we consider one particular model that is based on a flat spacetime with broken SO(2,3)$_\star$ symmetry [11,12].

Any physical model that includes noncommutative effects and that reduces to conventional physics in the proper limit is expected to break Lorentz symmetry [13]. A general framework for the study of Lorentz violation has been developed over the last 30 years [14–18]. Indeed, numerous experimental and observational limits exist already on many different a priori independent types of Lorentz violation [19]. Additionally, this effective-field-theory framework should contain any realistic noncommutative model. This has already been shown for non-gravitational models [13]. In this work, we argue that the noncommutative SO(2,3)$_\star$ gravity model also fits into the gravitational sector of the Standard-Model Extension (SME). This serves as an example of the general notion that the SME contains all specific action-based Lorentz-violating models.

2. Noncommutative SO(2,3)$_\star$ Gravity

Consider a model consisting of a flat four-dimensional spacetime with an SO(2,3) gauge field [12]. Suppose that this symmetry is spontaneously broken along a timelike direction, with the field in that direction achieving a vacuum expectation value ℓ. The corresponding action takes the form of a model of gravity, with pieces corresponding to Einstein–Hilbert terms, cosmological-constant terms, and Gauss-Bonnet terms; this action is symmetric under an SO(1,3) subgroup of the broken SO(2,3) symmetry. If conventional field products are then translated into Moyal–Weyl $\star$-products and a Seiberg–Witten map is used to re-express quantities in terms of commutative products, we get a broken-SO(2,3)$_\star$ gravitational theory.

This process has been carried out in Ref. [12] and we present the relevant results here. This theory may be expressed as a model with noncommutative local SO(1,3)$_\star$ symmetry. The result is expanded in terms of the noncommutative background $\theta^{\alpha\beta}$, with leading terms at second order in this quantity. To display the action, we note that the geometric quantities that will appear use conventional notation: $e_\mu{}^a$ is the vierbein (with determinant e), $\omega_\gamma{}^{ab}$ is the associated spin connection, $\Gamma^\rho{}_{\gamma\alpha}$ are the Christoffel symbols associated with spacetime metric $g_{\alpha\beta}$, $R_{\alpha\beta\gamma\delta}$ is the Riemann tensor, $R_{\alpha\beta}$ is the Ricci tensor, and R is the curvature scalar.

Once spacetime torsion $T_{\lambda\mu\nu}$ is set to zero, the action for the model ([12], Equation (4.2)) may be expressed in the form

$$
\begin{aligned}
S_{NCR} \;=\; & -\tfrac{1}{2\kappa} \int d^4x\, e \left[R - \tfrac{6}{\ell^2}(1 + c_2 + 2c_3) \right] \\
& + \tfrac{1}{16\kappa\ell^4} \int d^4x \sum_{u=1}^{6} e\,\theta^{\alpha\beta}\theta^{\gamma\delta} C_{(u)} L^{(u)}_{\alpha\beta\gamma\delta} \quad,
\end{aligned}
\tag{1}
$$

where $\kappa = 8\pi G_N$ and ℓ is a length parameter. The antisymmetric coefficients $\theta^{\alpha\beta}$ are to be thought of as a fixed background field describing the degree of noncommutativity of spacetime. Note that natural units are adopted ($\hbar = c = 1$), which implies that ℓ has units of length or inverse mass and θ has units of length squared.

The top row of Equation (1) is the action for conventional general relativity with a cosmological constant $\Lambda = -3\left(\frac{1 + c_2 + 2c_3}{\ell^2}\right)$. (Note that this is the correct value of the cosmological constant only in the commutative limit $\theta = 0$. For $\theta \neq 0$, other terms in the action will also effectively contribute to it.) The parameters c_2 and c_3 describe the relative weights of various contributions to the unbroken $SO(2,3)_\star$ action. Thus, the action S_{CNR} may be thought of as a family of actions parameterized by c_2, c_3, and ℓ. The tensors $L^{(u)}_{\alpha\beta\gamma\delta}$ are geometric quantities; the weights $C_{(u)}$ measure the relative contributions of these quantities to the action. The tensors and their weights are listed in Table 1.

Table 1. Geometric quantities and their weights that appear in the action S_{CNR}.

u	Weight $C_{(u)}$	Geometric Quantity $L^{(u)}_{\alpha\beta\gamma\delta}$
1	$3c_2 + 16c_3$	$R_{\alpha\beta\gamma\delta}$
2	$-6 - 22c_2 - 36c_3$	$g_{\beta\delta} R_{\alpha\gamma}$
3	$\frac{1}{\ell^2}(6 + 28c_2 + 56c_3)$	$g_{\alpha\gamma} g_{\beta\delta}$
4	$-4 - 16c_2 - 32c_3$	$e_a^\mu e_{\beta b}(\widetilde{\nabla}_\gamma e_\alpha^a)(\widetilde{\nabla}_\delta e_\mu^b)$
5	$4 + 12c_2 + 32c_3$	$e_{\delta a} e_b^\mu (\widetilde{\nabla}_\alpha e_\gamma^a)(\widetilde{\nabla}_\beta e_\mu^b)$
6	$2 + 4c_2 + 8c_3$	$g_{\beta\delta} e_a^\mu e_b^\nu [(\widetilde{\nabla}_\alpha e_\nu^a)(\widetilde{\nabla}_\gamma e_\mu^b) - (\widetilde{\nabla}_\gamma e_\mu^a)(\widetilde{\nabla}_\alpha e_\nu^b)]$

The adjusted covariant derivatives $\widetilde{\nabla}_\mu$ of the vierbein that appear in terms 4 through 6 include contributions from the $SO(1,3)$ connection but not from the Christoffel symbols:

$$
\widetilde{\nabla}_\gamma e_\alpha{}^a = \partial_\alpha e_\alpha{}^a + \omega_\gamma{}^{ab} e_{ab} = \nabla_\gamma e_\alpha{}^a + \Gamma^\rho{}_{\gamma\alpha} e_\rho{}^a \quad.
\tag{2}
$$

If the vierbein satisfies the usual compatibility condition $\nabla_\gamma e_\alpha{}^a = 0$, then the adjusted covariant derivative may be expressed as

$$
\widetilde{\nabla}_\gamma e_\alpha{}^a = \Gamma^\rho{}_{\gamma\alpha} e_\rho{}^a \quad.
\tag{3}
$$

This implies the explicit appearance of the Christoffel symbols in the Lagrangian, the consequences of which are discussed in the next section.

The model acts like a relativistic theory of gravity in several ways, but there are some issues with interpreting it as such. For example, it is derived with the assumption that $\partial_\alpha \theta^{\mu\nu} = 0$. This assumption is reasonable in the original flat-spacetime context of the model. However, if the model is to be interpreted in curved spacetime, this assumption is clearly coordinate dependent. We may attempt to fix this issue by instead assuming that $\nabla_\alpha \theta^{\mu\nu} = 0$, but even this condition cannot apply in many situations. Nonzero tensor fields with vanishing covariant derivative cannot exist on many manifolds, including, say, spacetime with a Schwarzschild metric [15,20]. Therefore, if we wish to seriously consider action 1 to represent a theory of gravity, then we must consider it to be an approximation to a more realistic model with $\nabla_\alpha \theta^{\mu\nu} \neq 0$. In what follows, we will assume that terms involving derivatives of $\theta^{\mu\nu}$ that may appear in a more-realistic model are negligible in comparison to all other terms.

3. Gravitational Sector of the Lorentz-Violating Standard-Model Extension

The full action [15] describing the gravitational sector of the SME can be expressed as a sum of terms, each of which contracts a coefficient with spacetime indices with geometric quantities such as the Riemann tensor $R_{\alpha\beta\gamma\delta}$, the torsion $T_{\lambda\mu\nu}$, and their covariant derivatives:

$$S_{\text{gravity}} = \frac{1}{2\kappa} \int d^4x \, e \left[(k_T)^{\lambda\mu\nu} T_{\lambda\mu\nu} + (k_R)^{\kappa\lambda\mu\nu} R_{\kappa\lambda\mu\nu} + (k_{DT})^{\kappa\lambda\mu\nu} D_\kappa T_{\lambda\mu\nu} + \cdots \right] \, . \tag{4}$$

The tensors k_T, k_R, etc. are coefficients for Lorentz and diffeomorphism violation and the ellipses represent terms with higher powers of curvature and torsion and derivative terms [21,22]. Note that a violation of local Lorentz symmetry generically implies a violation of diffeomorphism symmetry, as explained in the literature [15,23]. As with $\theta^{\mu\nu}$, it is not possible for the coefficients to be covariant derivative constants on most spacetime manifolds, and so they must be functions of spacetime position, though we may assume that their partial derivatives are negligible in experimentally relevant frames.

In this work, we consider two limits of this full action: the minimal set of terms necessary for Lorentz violation and the weakly-curved-spacetime limit (or quadratic limit) of the full action.

3.1. Covariant Match

In the gravity sector of the fully observer-covariant SME, the minimal set of terms that arises are given by the action [15],

$$S_{LV,\text{cov}} = \frac{1}{2\kappa} \int d^4x \, e \left[R + (k_R)_{\alpha\beta\gamma\delta} R^{\alpha\beta\gamma\delta} \right], \tag{5}$$

where $(k_R)_{\alpha\beta\gamma\delta}$ are the 20 (background) coefficients for local Lorentz and diffeomorphism violation. It is clear that there is overlap with the noncommutative model Equation (1). However, there are no terms in the SME containing explicit dependence on the non-tensorial connection coefficients $\Gamma^\alpha{}_{\beta\gamma}$.

It is important at this stage to distinguish two types of symmetry transformations. The first is called an *observer* diffeomorphism, or general coordinate transformation, which is a diffeomorphism that affects both the background, $(k_R)_{\alpha\beta\gamma\delta}$ and the dynamical fields $e_\mu{}^a$. The second is called a *particle* diffeomorphism, which is a diffeomorphism that leaves the background $(k_R)_{\alpha\beta\gamma\delta}$ unchanged while the dynamical fields $e_\mu{}^a$ transform in the usual way. It is this second type of symmetry breaking, *particle* diffeomorphism symmetry breaking, that is described by the SME approach and is broken by the second term in Equation (5). Because the action terms in the SME are scalars under general coordinate transformations, they trivially satisfy observer symmetry. These points are discussed in more detail in the literature [15,24,25].

The explicit appearance of $\Gamma^\rho{}_{\gamma\alpha}$ in terms 4–6 of Equation (1) implies that each of these terms is not symmetric under observer diffeomorphisms. Whether the model can be massaged into an observer covariant form, for example by a special choice of the parameters c_2 and c_3, remains to be shown. Note that the model does appear covariant under *observer* local Lorentz transformations while breaking *particle* local Lorentz symmetry. Despite the difficulty, we can proceed at the quadratic-action level, where a model that breaks observer diffeomorphism invariance cannot be distinguished from a model that breaks particle diffeomorphism invariance.

3.2. Linearized Lorentz-Violating Standard-Model Extension

If we restrict the full SME to a version with equations of motion that are linear in $h_{\mu\nu}$ [26–28], then the action takes the form, after a rescaling by $1/2\kappa$,

$$
S = \int d^4x \left[\mathcal{L}_0 + \frac{1}{8\kappa} h_{\mu\nu} \sum_d \widehat{\mathcal{K}}^{(d)\mu\nu\rho\sigma} h_{\rho\sigma} \right] \; .
\tag{6}
$$

In this expression, $\mathcal{L}_0 = e(R - 2\Lambda)/2\kappa$ is the usual quadratic Einstein–Hilbert Lagrange density and $h_{\mu\nu} := g_{\mu\nu} - \eta_{\mu\nu}$ is the metric perturbation, assumed to be small. The $\widehat{\mathcal{K}}^{(d)\mu\nu\rho\sigma}$ are general derivative operators formed from background coefficients and derivatives. The summation is over the mass dimension d of the operators. In general, apart from surface terms, this sum includes 14 classes of irreducible representations involving tensors and derivative operators, all detailed in [28].

A primary goal of this paper is to argue that the non-commutative (broken-)SO(2,3)$_\star$ action S_{NCR} in the linearized limit is a special case of this general linearized Lorentz-violating action. We explicitly calculate the map that shows this correspondence. We will show that the subset of the operator terms in the action Equation (6) that occur in the non-commutative model Equation (1) can be written as

$$
S_{LV,NC} = \frac{1}{8\kappa} \int d^4x\, h_{\mu\nu} \Big\{ \big[s^{(4)\mu\rho\alpha\nu\sigma\beta} + s^{(4,1)\mu\rho\nu\sigma\alpha\beta} + s^{(4,2)\mu\rho\alpha\nu\sigma\beta} + k^{(4,3)\mu\alpha\nu\beta\rho\sigma} \big] \partial_\alpha \partial_\beta \\
+ s^{(2,1)\mu\rho\nu\sigma} + k^{(2,1)\mu\nu\rho\sigma} \Big\} h_{\rho\sigma}.
\tag{7}
$$

Each of these terms has distinct tensor symmetries described by a particular Young tableau [29]. The coefficients with the $(4, \#)$ label are coefficients for mass dimension 4 operators, while those without derivatives labeled $(2, \#)$ are coefficients for mass dimension 2 operators. The latter represent an arbitrary mass matrix for the gravitational fluctuations $h_{\mu\nu}$. Incidently, none of the terms in Equation (1) contain odd mass dimension operators, and therefore the CPT symmetry is maintained.

4. Connecting NC SO(3,2)$_\star$ Gravity to the SME

The action for any linearized theory of gravity is quadratic in the perturbation $h_{\mu\nu}$. Therefore, we need to calculate each of the quantities that appears in S_{NCR} to second order in $h_{\mu\nu}$. Calculations of these quantities to first order are widespread in the literature, but calculations to second order are not, so we summarize the key results in the Appendix. With these formulæ, we may expand the noncommutative action S_{NCR} in powers of $h_{\mu\nu}$. The results may then be manipulated into the form of the linearized action Equation (6).

First, we show the match to the SME for the *massive* $u = 3$ term. Expanding this term from Equation (1) in the quadratic action limit, we obtain

$$
\begin{aligned}
S_{NC,Mass} &= \frac{C_{(3)}}{16\kappa\ell^6} \int d^4x\, e\,\theta^{\alpha\beta}\theta^{\gamma\delta} g_{\alpha\gamma} g_{\beta\delta} \\
&= \frac{1}{8\kappa} \int d^4x \Big\{ \frac{C_{(3)}}{2\ell^6} \Big(\theta^2 + \big[\tfrac{1}{2}\theta^2 \eta^{\mu\nu} + 2\theta_\alpha{}^\mu \theta^{\alpha\mu} \big] h_{\mu\nu} \Big) \\
&\qquad + \frac{C_{(3)}}{16\ell^4} h_{\mu\nu} \big[\theta^2 \eta^{\mu\nu}\eta^{\rho\sigma} - 2\theta^2 \eta^{\mu\rho}\eta^{\nu\sigma} + 8\theta_\alpha{}^\mu \theta^{\alpha\nu}\eta^{\rho\sigma} + 8\theta^{\mu\rho}\theta^{\nu\sigma} \big] h_{\rho\sigma} \Big\},
\end{aligned}
\tag{8}
$$

where $\theta^2 := \theta_{\mu\nu}\theta^{\mu\nu}$. Note that all indices on the right-hand sides of these expressions are raised and lowered with η, as they are considered to act in the flat spacetime with field $h_{\mu\nu}$. The first term with just θ^2 is a constant and irrelevant for dynamics, while the second term linear in $h_{\mu\nu}$ acts as a constant contribution to the stress-energy tensor (of the form of a cosmological constant). The last line can be matched to the last two terms in Equation (7) using Young tableau projections. The coefficients appearing, $s^{(2,1)\mu\rho\nu\sigma}$ and $k^{(2,1)\mu\nu\rho\sigma}$, correspond to the Young tableaus $\begin{smallmatrix}\mu & \nu \\ \rho & \sigma\end{smallmatrix}$ and $\boxed{\mu\,\nu\,\rho\,\sigma}$, respectively. The explicit results we find are

$$
\begin{aligned}
s^{(2,1)\mu\rho\nu\sigma} &= \frac{C_{(3)}}{12\ell^4}\Big[2\eta^{\mu\nu}\theta^{\rho\alpha}\theta^{\sigma}{}_{\alpha} + 2\eta^{\rho\sigma}\theta^{\mu\alpha}\theta^{\nu}{}_{\alpha} - 2\eta^{\rho\nu}\theta^{\sigma\alpha}\theta^{\mu}{}_{\alpha} - 2\eta^{\mu\sigma}\theta^{\rho\alpha}\theta^{\nu}{}_{\alpha} \\
&\qquad + 2\theta^{\rho\nu}\theta^{\sigma\mu} + 4\theta^{\rho\mu}\theta^{\sigma\nu} + 2\theta^{\mu\nu}\theta^{\rho\sigma} + (\eta^{\rho\sigma}\eta^{\mu\nu} - \eta^{\rho\nu}\eta^{\sigma\mu})\,\theta^2\Big] \quad, \\
k^{(2,1)\mu\nu\rho\sigma} &= \frac{C_{(3)}}{48\ell^4}\Big[4\eta^{\mu\nu}\theta^{\rho\alpha}\theta^{\sigma}{}_{\alpha} + 4\eta^{\rho\sigma}\theta^{\mu\alpha}\theta^{\nu}{}_{\alpha} + 4\eta^{\rho\nu}\theta^{\sigma\alpha}\theta^{\mu}{}_{\alpha} + 4\eta^{\mu\sigma}\theta^{\rho\alpha}\theta^{\nu}{}_{\alpha} \\
&\qquad + 4\eta^{\sigma\nu}\theta^{\rho\alpha}\theta^{\mu}{}_{\alpha} + 4\eta^{\rho\mu}\theta^{\sigma\alpha}\theta^{\nu}{}_{\alpha} - (\eta^{\rho\sigma}\eta^{\mu\nu} + \eta^{\rho\nu}\eta^{\sigma\mu} + \eta^{\rho\mu}\eta^{\sigma\nu})\,\theta^2\Big] \quad.
\end{aligned}
\tag{9}
$$

We classify the remaining terms in Equation (1) as *kinetic* terms that only involve mass dimension 4 operators. After expanding these terms in the quadratic-action limit and manipulating the result into the form of Equation (7), we obtain

$$
S_{\text{NC,Kin}} = \frac{1}{8\kappa}\int d^4x\, h_{\mu\nu}(K_{\text{NC}})^{\mu\nu\rho\sigma\alpha\beta}\partial_\alpha\partial_\beta h_{\rho\sigma},
\tag{10}
$$

where the quantity $(K_{\text{NC}})^{\mu\nu\rho\sigma\alpha\beta}$ is given by

$$
\begin{aligned}
(K_{\text{NC}})^{\mu\nu\rho\sigma\alpha\beta} &= \tfrac{1}{16\ell^4}(2C_{(1)} - 2C_{(2)} + C_{(4)})(\eta^{\alpha\beta}\theta^{\rho\nu}\theta^{\sigma\mu} + \eta^{\alpha\beta}\theta^{\rho\mu}\theta^{\sigma\nu}) \\
&\quad + \tfrac{1}{64\ell^4}(4C_{(1)} - 2C_{(2)} + C_{(4)})\big(\{(\eta^{\nu\alpha}\theta^{\beta\rho}\theta^{\sigma\mu} - \eta^{\sigma\alpha}\theta^{\beta\mu}\theta^{\rho\nu} + \eta^{\nu\beta}\theta^{\alpha\rho}\theta^{\sigma\mu} - \eta^{\sigma\beta}\theta^{\alpha\mu}\theta^{\rho\nu}) \\
&\qquad\qquad + (\rho \rightleftharpoons \sigma)\} + \{\mu \rightleftharpoons \nu\}\big) \\
&\quad + \tfrac{1}{16\ell^4}(2C_{(1)} + C_{(2)} - C_{(5)})(\eta^{\mu\nu}\theta^{\rho\alpha}\theta^{\beta\sigma} + \eta^{\rho\sigma}\theta^{\mu\alpha}\theta^{\beta\nu} + \eta^{\mu\nu}\theta^{\rho\beta}\theta^{\alpha\sigma} + \eta^{\rho\sigma}\theta^{\mu\beta}\theta^{\alpha\nu}) \\
&\quad + \tfrac{1}{16\ell^4}(C_{(2)} - C_{(6)})\big(\{(\tfrac{1}{2}\eta^{\sigma\alpha}\eta^{\beta\nu}\theta^{\rho\gamma}\theta^{\mu}{}_{\gamma} + \tfrac{1}{2}\eta^{\sigma\beta}\eta^{\alpha\nu}\theta^{\rho\gamma}\theta^{\mu}{}_{\gamma} - \eta^{\sigma\nu}\eta^{\alpha\beta}\theta^{\rho\gamma}\theta^{\mu}{}_{\gamma}) + (\mu \rightleftharpoons \nu) \\
&\qquad\qquad + \eta^{\rho\nu}\eta^{\sigma\mu}\theta^{\alpha\gamma}\theta^{\beta}{}_{\gamma}\} + \{\rho \rightleftharpoons \sigma\} - 2\eta^{\rho\sigma}\eta^{\mu\nu}\theta^{\alpha\gamma}\theta^{\beta}{}_{\gamma}\big) \\
&\quad + \tfrac{1}{16\ell^4}C_{(1)}\big(\{(\eta^{\sigma\nu}\theta^{\rho\alpha}\theta^{\beta\mu} + \eta^{\sigma\nu}\theta^{\rho\beta}\theta^{\alpha\mu}) + (\mu \rightleftharpoons \nu)\} + \{\rho \rightleftharpoons \sigma\}\big) \quad.
\end{aligned}
\tag{11}
$$

At this stage, one can project Equation (11) into the irreducible tensors that appear in Equation (7). Consider the first coefficients, $s^{(4)\mu\rho\alpha\nu\sigma\beta}$, for which the operator it is contracted with, $\sim h\partial\partial h$, is a gauge invariant combination (invariant under the transformation $\delta h_{\mu\nu} = -\partial_\mu\xi_\nu - \partial_\nu\xi_\mu$). Calculation with Young Tableau projection P_Y reveals

$$
\begin{aligned}
s^{(4)\mu\rho\alpha\nu\sigma\beta} &= P_Y^{\begin{smallmatrix}\mu & \nu \\ \rho & \sigma \\ \alpha & \beta\end{smallmatrix}}(K_{\text{NC}})^{\mu\nu\rho\sigma\alpha\beta} \\
&= \frac{1}{36\ell^4}(2C_{(1)} - 3C_{(2)} + C_{(4)} + C_{(5)})\big(\tfrac{1}{2}\eta^{\rho\sigma}\theta^{\alpha\nu}\theta^{\beta\mu} - \tfrac{1}{2}\eta^{\nu\rho}\theta^{\alpha\sigma}\theta^{\beta\mu} + \eta^{\rho\sigma}\theta^{\alpha\mu}\theta^{\beta\nu} + ...\big),
\end{aligned}
\tag{12}
$$

where the ellipses stand for the remaining symmetrizing terms. The explicit terms are not shown for brevity and because this contribution can be more profitably expressed using an equivalent two-tensor set of coefficients defined by

$$
\bar{s}_{\gamma\delta} = -\tfrac{1}{36}\epsilon_{\mu\alpha\gamma}\epsilon_{\nu\sigma\beta\delta}s^{(4)\mu\rho\alpha\nu\sigma\beta}.
\tag{13}
$$

Employing this, the portion of the Lagrangian containing the $s^{(4)}$ coefficients can be expressed as

$$
\mathcal{L}_{\text{LV,NC}} \supset \tfrac{1}{4\kappa}\int d^4x\, h_{\mu\nu}\bar{s}_{\kappa\lambda}\mathcal{G}^{\mu\kappa\nu\lambda},
\tag{14}
$$

where, for the non-commutative model under study, we have

$$
\bar{s}_{\kappa\lambda} = -\tfrac{1}{24\ell^4}(2C_{(1)} - 3C_{(2)} + C_{(4)} + C_{(5)})\left(\theta_{\kappa\alpha}\theta_\lambda{}^{\alpha} - \tfrac{1}{4}\eta_{\kappa\lambda}\theta^2\right),
\tag{15}
$$

and we have removed the trace of these coefficients since they contribute only as a scaling of GR at this level. This result shows that the non-commutative model overlaps with, in part, the minimal SME gravity sector in the weak-field limit. In this model, the nine coefficients $\bar{s}_{\kappa\lambda}$ are evidently controlled by the six non-commutative parameters $\theta^{\alpha\beta}$. Note that the size of these coefficients depends on the relative size of the non-commutative parameters and the the length parameter ℓ.

For the other classes of coefficients appearing in Equation (7), we can proceed in a similar fashion with the Young Tableau projection. All terms are summarized in Table 2 below. The explicit expressions for the Young projections are lengthy and omitted here for brevity, but they can be calculated with standard methods [29].

Table 2. Young Projections for the *kinetic* portion of the NC action.

SME Coefficients	Young Projection
$s^{(4)\mu\rho\alpha\nu\sigma\beta}$	$P_Y{\tiny\begin{array}{\|c\|c\|}\hline \mu & \nu \\\hline \rho & \sigma \\\hline \alpha & \beta \\\hline\end{array}}(K_{\mathrm{NC}})^{\mu\nu\rho\sigma\alpha\beta}$
$s^{(4,1)\mu\rho\nu\sigma\alpha\beta}$	$P_Y{\tiny\begin{array}{\|c\|c\|c\|c\|}\hline \mu & \nu & \alpha & \beta \\\hline \rho & \sigma \\\cline{1-2}\end{array}}(K_{\mathrm{NC}})^{\mu\nu\rho\sigma\alpha\beta}$
$s^{(4,2)\mu\rho\alpha\nu\sigma\beta}$	$P_Y{\tiny\begin{array}{\|c\|c\|c\|}\hline \mu & \nu & \beta \\\hline \rho & \sigma \\\cline{1-2}\alpha \\\cline{1-1}\end{array}}(K_{\mathrm{NC}})^{\mu\nu\rho\sigma\alpha\beta}$
$k^{(4,3)\mu\alpha\nu\beta\rho\sigma}$	$P_Y{\tiny\begin{array}{\|c\|c\|c\|c\|}\hline \mu & \nu & \rho & \sigma \\\hline \alpha & \beta \\\cline{1-2}\end{array}}(K_{\mathrm{NC}})^{\mu\nu\rho\sigma\alpha\beta}$

5. Conclusions, Prospects for Further Work

We have shown that the model proposed in Ref. [12], in its quadratic limit, is a subset of the Lorentz- and diffeomorphism-violating Standard-Model Extension. The main results are understood as a series of Young Tableau maps described in Section 4.

One consequence of the match obtained relates to experimental and observational constraints on the noncommutative model considered. For the gauge-preserving portion of the Lagrangian, for which the observable effects are controlled by the $\bar{s}_{\mu\nu}$ coefficients, an extensive study of phenomenology has been performed [30–34]. To date, numerous experiments and observations have reported measurements on these coefficients [19,35,36]. The best current astrophysical limits come from a recent comparison of the arrival times of electromagnetic and gravitational waves from a pair of colliding neutron stars [37]. Lunar laser ranging and ground-based gravimetry also place limits on these coefficients [38–41]. The best limits imply constraints on the order of $\bar{s}_{\mu\nu} < 10^{-14}$. Heuristically then, this would imply that the non-commutivity coefficients $\theta^{\alpha\beta}$ and the length parameter ℓ are related by $\theta^2/\ell^4 < 10^{-15}$. However, a more precise statement would require a thorough phenomenological analysis of the diffeomorphism-violating terms in Section 4 above.

It would be of interest to explore the role of additional terms in the non-commutative model, as in Ref. [11] that involve higher derivatives. These terms have been generally classified in the SME approach and a match should exist [28].

Author Contributions: Both authors contributed equally to the research and writing of this work.

Funding: Q.G.B. acknowledges support from the National Science Foundation under Grant No. PHY-1806871.

Acknowledgments: Thanks to Berry College and the Indiana University Center for Spacetime Symmetries for financial support during the creation of this work.

Conflicts of Interest: The author declares no conflict of interest.

Appendix A. Geometric Quantities to 2nd Order in the Metric Perturbation

Consider a pair of theories. The first operates in curved-spacetime, including a manifold $\mathcal{M}$, a metric $g_{\mu\nu}$, a local flat metric for tangent spaces η_{ab}, and a set of vierbein $e_\mu{}^a$ that relate the metrics through $g_{\mu\nu} = e_\mu{}^a e_\nu{}^b \eta_{ab}$. (Equivalently, the vierbein may be thought of as a position-dependent change-of-basis matrix that relates a manifold coordinate basis $\{\vec{v}_\mu\}$ to a local tangent-space basis $\{\vec{u}_a\}$.) The second theory operates in a flat spacetime with an auxiliary field $h_{\mu\nu}$. For this theory, the manifold is simply $\mathbb{R}^4$, the manifold metric is $\eta_{\mu\nu}$, the tangent-space metric is η_{ab}, and global coordinates may be found so that the vierbein is just the Kronecker delta $\delta_\mu{}^a$.

A perturbation scheme is a map

$$(\mathcal{M}, g_{\mu\nu}, \eta_{ab}, e_\mu{}^a) \to (\mathbb{R}^4, \eta_{\mu\nu}, \eta_{ab}, \delta_\mu{}^a) + h_{\mu\nu} \tag{A1}$$

between these theories so that they approximately describe the same physical effects. In particular, we will consider situations where $g_{\mu\nu} \approx \eta_{\mu\nu}$, so that the map may be nicely approximated by a power series in $g_{\mu\nu} - \eta_{\mu\nu}$. We wish to calculate an action in terms of $h_{\mu\nu}$ that mimics the physical effects of the original theory up to order h^2.

The first piece of the map is defined by the correspondence

$$g_{\mu\nu} = \eta_{\mu\nu} + h_{\mu\nu} \quad . \tag{A2}$$

This is the definition of $h_{\mu\nu}$ and hence is correct to all orders in h. Our goal in this section is to find expressions for other geometric quantities $g^{\mu\nu}$, $e_\mu{}^a$, and so on that appear in the action of the full theory. The formulas for these quantities should only involve the flat-spacetime tensors $h_{\mu\nu}$, $\eta_{\mu\nu}$, η_{ab}, and $\delta_\mu{}^a$.

It is important to note that the defining map Equation (A2) is not a tensor equation in the original spacetime. This implies that indices on $h_{\mu\nu}$ cannot be raised and lowered like the indices of true tensors. That is, $h^{\mu\nu}$ is *not* equal to $g^{\mu\alpha} g^{\nu\beta} h_{\alpha\beta}$. The geometry of the original manifold does not by itself define a unique value of such quantities, and we have some freedom in choosing our definition of them. The most convenient choice is defining them so that $h_{\mu\nu}$ acts like a true tensor in the flat spacetime. That is, we pick $h^\mu{}_\nu := \eta^{\mu\alpha} h_{\alpha\nu}$, $h^{\mu\nu} := \eta^{\mu\alpha} \eta^{\nu\beta} h_{\alpha\beta}$, etc. Similarly, we may choose to relate global and tangent-space indices with the flat-space vierbein $\delta^\mu{}_a$: $h_{\mu a} := h_{\mu\nu} \delta^\nu{}_a$, $h_\mu{}^a := h_{\mu\nu} \eta^{\nu\lambda} \delta^a{}_\lambda$, etc.

The raised-index metric $g^{\mu\nu}$ may then be evaluated to second order in $h_{\mu\nu}$ through the following strategy. The fundamental definition of $g^{\mu\nu}$ is that it is the matrix inverse of $g_{\mu\nu}$:

$$\delta_\mu{}^\lambda = g_{\mu\nu} g^{\nu\lambda} \quad . \tag{A3}$$

We proceed by using the ansatz $g^{\nu\lambda} = \eta^{\nu\lambda} + j^{\nu\lambda} + k^{\nu\lambda} + o(h^3)$ where $j^{\nu\lambda}$ is first order in h and $k^{\nu\lambda}$ is second order. If we insist that Equation (A3) hold order-by-order in h, then we need

$$j^{\alpha\lambda} = -\eta^{\alpha\mu} \eta^{\nu\lambda} h_{\mu\nu} \quad \text{and} \quad k^{\alpha\lambda} = \eta^{\alpha\mu} \eta^{\nu\beta} \eta^{\lambda\gamma} h_{\mu\nu} h_{\beta\gamma} \quad . \tag{A4}$$

Using the definitions of upper-index h quantities described in the previous paragraph, we may then write

$$g^{\mu\nu} = \eta^{\mu\nu} - h^{\mu\nu} + h^{\mu\alpha} h_\alpha{}^\nu + o(h^3) \quad . \tag{A5}$$

Note again that $g^{\mu\nu} \neq \eta^{\mu\nu} + h^{\mu\nu}$ as the breakdown of $g_{\mu\nu}$ into $\eta_{\mu\nu} + h_{\mu\nu}$ is not a true tensor operation.

The quadratic approximation for the vierbein may be calculated by using the ansatz $e_\mu{}^a = \delta_\mu{}^a + f_\mu{}^a + \ell_\mu{}^a + o(h^3)$, where f is first order in h and ℓ is second order, and insisting that the exact relation

$$g_{\mu\nu} = e_\mu{}^a e_\nu{}^b \eta_{ab} \tag{A6}$$

hold order-by-order in h. This results in the expression

$$e_\mu{}^a = \delta_\mu{}^a + \tfrac{1}{2} h_\mu{}^a - \tfrac{1}{8} h_{\mu\lambda} h^{\lambda a} + o(h^3) \quad , \tag{A7}$$

where again h quantities are related to each other with the flat-spacetime metrics $\eta_{\mu\nu}, \eta_{ab}$ and flat-spacetime vierbein $\delta_\mu{}^a$. Explicitly, $h_\mu{}^a := \eta^{\nu\rho}\delta_\rho{}^a h_{\mu\nu}$ and $h^{\lambda a} := \eta^{\lambda\mu}\eta^{\nu\rho}\delta_\rho{}^a h_{\mu\nu}$.

Once we have these, calculations of other geometric quantities are rather straightforward if tedious.

Metric:

$$\begin{aligned}
g_{\mu\nu} &= \eta_{\mu\nu} + h_{\mu\nu} \quad , \\
g^{\mu\nu} &= \eta^{\mu\nu} - h^{\mu\nu} + h^{\mu\alpha}h_\alpha{}^\nu + o(h^3) \quad .
\end{aligned} \tag{A8}$$

Vierbein:

$$\begin{aligned}
e_\mu{}^a &= \delta_\mu{}^a + \tfrac{1}{2}h_\mu{}^a - \tfrac{1}{8}h_{\mu\lambda}h^{\lambda a} + o(h^3) \quad , \\
e_{\mu a} &= \eta_{\mu a} + \tfrac{1}{2}h_{\mu a} - \tfrac{1}{8}h_{\mu\lambda}h^\lambda{}_a + o(h^3) \quad , \\
e^{\mu a} &= \eta^{\mu a} - \tfrac{1}{2}h^{\mu a} + \tfrac{3}{8}h^\mu{}_\lambda h^{\lambda a} + o(h^3) \quad , \\
e^\mu{}_a &= \delta^\mu{}_a - \tfrac{1}{2}h^\mu{}_a + \tfrac{3}{8}h^{\mu\lambda}h_{\lambda a} + o(h^3) \quad , \\
e := \det(e_\mu{}^a) &= 1 + \tfrac{1}{2}h_\mu{}^\mu + \tfrac{1}{8}\left(h_\mu{}^\mu h_\nu{}^\nu - 2h_\mu{}^\nu h_\nu{}^\mu\right) + o(h^3) \quad .
\end{aligned} \tag{A9}$$

Note again that the expressions for the vierbein quantities cannot be related to each other simply by raising and lowering indices: $e^{\mu a} \neq \eta^{\mu\lambda}e_\lambda{}^a$, etc. Note also that the index placement in the definition of e is important: $\det(e^\mu{}_a) = \dfrac{1}{\det(e_\mu{}^a)}$.

Connection coefficients:

$$\begin{aligned}
\Gamma_{\alpha\mu\nu} &= \tfrac{1}{2}(\partial_\mu h_{\nu\alpha} + \partial_\nu h_{\mu\alpha} - \partial_\alpha h_{\mu\nu}) \quad , \\
\Gamma^\alpha{}_{\mu\nu} &= \tfrac{1}{2}(\eta^{\alpha\sigma} - h^{\alpha\sigma})(\partial_\mu h_{\nu\sigma} + \partial_\nu h_{\mu\sigma} - \partial_\sigma h_{\mu\nu}) + o(h^3) \quad , \\
\omega_\mu{}^{ab} &= \left[-\tfrac{1}{2}\partial^a h_\mu{}^b + -\tfrac{1}{8}h^{a\lambda}\partial_\mu h_\lambda{}^b + \tfrac{1}{4}h^{a\lambda}\partial_\lambda h_\mu{}^b - \tfrac{1}{4}h^{a\lambda}\partial^b h_{\lambda\mu}\right] - \left[a \rightleftharpoons b\right] + o(h^3) \quad .
\end{aligned} \tag{A10}$$

Derivative compatibility:

$$\begin{aligned}
\nabla_\gamma g_{\mu\nu} &= 0 \quad , \\
\nabla_\gamma e_\mu{}^a &= 0 \quad .
\end{aligned} \tag{A11}$$

Riemann tensor:

$$\begin{aligned}
R_{\alpha\beta\mu\nu} = \Big[&\Big(-\tfrac{1}{2}\partial_\alpha\partial_\mu h_{\beta\nu} - \tfrac{1}{8}\partial_\alpha h_{\mu\lambda}\partial_\beta h_\nu{}^\lambda - \tfrac{1}{8}\partial_\mu h_{\alpha\lambda}\partial_\nu h_\beta{}^\lambda - \tfrac{1}{8}\partial_\lambda h_{\alpha\mu}\partial^\lambda h_{\beta\nu} \\
&- \tfrac{1}{4}\partial_\alpha h_{\mu\lambda}\partial_\nu h_\beta{}^\lambda + \tfrac{1}{4}\partial_\alpha h_{\mu\lambda}\partial^\lambda h_{\beta\nu} + \tfrac{1}{4}\partial_\mu h_{\alpha\lambda}\partial^\lambda h_{\beta\nu}\Big) - \Big(\alpha \rightleftharpoons \beta\Big)\Big] - \Big[\mu \rightleftharpoons \nu\Big] + o(h^3).
\end{aligned} \tag{A12}$$

Ricci tensor:

$$\begin{aligned}
R_{\alpha\mu} = g^{\beta\nu}R_{\alpha\beta\mu\nu} = \Big[&\tfrac{1}{2}\partial_\alpha\partial_\lambda h_\mu{}^\lambda - \tfrac{1}{4}\partial_\alpha\partial_\mu h_\lambda{}^\lambda - \tfrac{1}{4}\partial_\lambda\partial^\lambda h_{\alpha\mu} \\
&- \tfrac{1}{2}h^{\lambda\rho}\Big(\partial_\alpha\partial_\lambda h_{\mu\rho} - \tfrac{1}{2}\partial_\alpha\partial_\mu h_{\lambda\rho} - \tfrac{1}{2}\partial_\lambda\partial_\rho h_{\alpha\mu}\Big) \\
&+ \Big(\tfrac{1}{4}\partial_\lambda h_\rho{}^\rho - \tfrac{1}{2}\partial^\rho h_{\rho\lambda}\Big)\Big(\partial_\alpha h_\mu{}^\lambda - \tfrac{1}{2}\partial^\lambda h_{\alpha\mu}\Big) \\
&- \tfrac{1}{4}(\partial_\lambda h_\mu{}^\rho)\Big(\partial_\rho h_\alpha{}^\lambda - \tfrac{1}{2}\partial^\lambda h_{\alpha\rho}\Big) + \tfrac{1}{8}(\partial_\alpha h_{\lambda\rho})(\partial_\mu h^{\lambda\rho})\Big] + \Big[\alpha \rightleftharpoons \mu\Big] + o(h^3).
\end{aligned} \tag{A13}$$

References

1. Snyder, H.S. Quantized space-time. *Phys. Rev.* **1947**, *71*, 38. [CrossRef]
2. Connes, A. *Noncommutative Geometry*; Academic Press: Cambridge, MA, USA, 1994.
3. Seiberg, N.; Witten, E. String theory and noncommutative geometry. *JHEP* **1999**, *9909*, 032. [CrossRef]
4. Aschieri, P.; Blohmann, C.; Dimitrijevic, M.; Meyer, F.; Schupp, P.; Wess, J. A Gravity theory on noncommutative spaces. *Class. Quant. Grav.* **2005**, *22*, 3511. [CrossRef]
5. Aschieri, P.; Dimitrijevic, M.; Meyer, F.; Wess, J. Noncommutative geometry and gravity. *Class. Quant. Grav.* **2006**, *23*, 1883. [CrossRef]
6. Ohl, T.; Schenkel, A. Cosmological and Black Hole Spacetimes in Twisted Noncommutative Gravity. *JHEP* **2009**, *0910*, 052. [CrossRef]
7. Yang, H.S. Emergent Gravity from Noncommutative Spacetime. *Int. J. Mod. Phys. A* **2009**, *24*, 4473. [CrossRef]

8. Steinacker, H. Emergent Geometry and Gravity from Matrix Models: an Introduction. *Class. Quant. Grav.* **2010**, *27*, 133001. [CrossRef]

9. Chamseddine, A.H. Deforming Einstein's gravity. *Phys. Lett. B* **2001**, *504*, 33. [CrossRef]

10. Cardella, M.A.; Zanon, D. Noncommutative deformation of four-dimensional Einstein gravity. *Class. Quant. Grav.* **2003**, *20*, L95. [CrossRef]

11. Ćirić, M.D.; Radovanović, V. Noncommutative $SO(2,3)_\star$ gauge theory and noncommutative gravity. *Phys. Rev. D* **2014**, *89*, 125021. [CrossRef]

12. Ćirić, M.D.; Nikolić, B.; Radovanović, V. Noncommutative $SO(2,3)_\star$ gravity: Noncommutativity as a source of curvature and torsion. *Phys. Rev. D* **2017**, *96*, 064029. [CrossRef]

13. Carroll, S.M.; Harvey, J.A.; Kostelecký, V.A.; Lane, C.D.; Okamoto, T. Noncommutative field theory and Lorentz violation. *Phys. Rev. Lett.* **2001**, *87*, 141601,

14. Kostelecký, V.A.; Samuel, S. Spontaneous Breaking of Lorentz Symmetry in String Theory. *Phys. Rev. D* **1989**, *39*, 683. [CrossRef]

15. Kostelecký, V.A. Gravity, Lorentz violation, and the standard model. *Phys. Rev. D* **2004**, *69*, 105009. [CrossRef]

16. Kostelecký, V.A.; Potting, R. CPT, strings, and meson factories. *Phys. Rev. D* **1995**, *51*, 3923. [CrossRef]

17. Colladay, D.; Kostelecký, V.A. CPT violation and the standard model. *Phys. Rev. D* **1997**, *55*, 6760. [CrossRef]

18. Colladay, D.; Kostelecký, V.A. Lorentz violating extension of the standard model. *Phys. Rev. D* **1998**, *58*, 116002. [CrossRef]

19. Kostelecký, V.A.; Russell, N. Data Tables for Lorentz and CPT Violation. *Rev. Mod. Phys.* **2011**, *83*, 11. [CrossRef]

20. Lane, C.D. Spacetime variation of Lorentz-violation coefficients at a nonrelativistic scale. *Phys. Rev. D* **2016**, *94*, 025016. [CrossRef]

21. Bailey, Q.G.; Kostelecký, V.A.; Xu, R. Short-range gravity and Lorentz violation. *Phys. Rev. D* **2015**, *91*, 022006. [CrossRef]

22. Bailey, Q.G. Anisotropic cubic curvature couplings. *Phys. Rev. D* **2016**, *94*, 065029. [CrossRef]

23. Bluhm, R.; Kostelecký, V.A. Spontaneous Lorentz Violation, Nambu-Goldstone Modes, and Gravity. *Phys. Rev. D* **2005**, *71*, 065008. [CrossRef]

24. Bluhm, R.; Fung, S.H.; Kostelecký, V.A. Spontaneous Lorentz and diffeomorphism violation, massive modes, and gravity. *Phys. Rev. D* **2008**, *77*, 065020. [CrossRef]

25. Bluhm, R. Explicit versus spontaneous diffeomorphism breaking in gravity. *Phys. Rev. D* **2015**, *91*, 065034. [CrossRef]

26. Kostelecký, V.A.; Tasson, J. Constraints on Lorentz violation from gravitational Cherenkov radiation. *Phys. Lett. B* **2015**, *749*, 551. [CrossRef]

27. Kostelecký, V.A.; Mewes, M. Testing local Lorentz invariance with gravitational waves. *Phys. Lett. B* **2016**, *757*, 510. [CrossRef]

28. Kostelecký, V.A.; Mewes, M. Lorentz and Diffeomorphism Violations in Linearized Gravity. *Phys. Lett. B* **2018**, *779*, 136. [CrossRef]

29. Hamermesh, M. *Group Theory and Its Application to Physical Problems*; Addison-Wesley: Reading, MA, USA, 1962.

30. Bailey, Q.G.; Kostelecký, V.A. Signals for Lorentz Violation in Post-Newtonian Gravity. *Phys. Rev. D* **2006**, *74*, 045001. [CrossRef]

31. Bailey, Q.G. Time delay and Doppler tests of the Lorentz symmetry of gravity. *Phys. Rev. D* **2009**, *80*, 044004, doi: 10.1103/PhysRevD.80.044004. [CrossRef]

32. Bailey, Q.G.; Tso, R. Light-bending tests of Lorentz invariance. *Phys. Rev. D* **2011**, *84*, 085025. [CrossRef]

33. Bailey, Q.G.; Everett, R.D.; Overduin, J.M. Limits on violations of Lorentz Symmetry from Gravity Probe B. *Phys. Rev. D* **2013**, *88*, 102001. [CrossRef]

34. Hees, A.; Bailey, Q.G.; Le Poncin-Lafitte, C.; Bourgoin, A.; Rivoldini, A.; Lamine, B.; Meynadier, F.; Guerlin, C.; Wolf, P. Testing Lorentz symmetry with planetary orbital dynamics. *Phys. Rev. D* **2015**, *92*, 064049. [CrossRef]

35. Hees, A.; Bailey, Q.G.; Bourgoin, A.; Bars, H.P.L.; Guerlin, C.; Poncin-Lafitte, C. Le. Tests of Lorentz symmetry in the gravitational sector. *Universe* **2016**, *2*, 4. [CrossRef]

36. Tasson, J. The Standard-Model Extension and Gravitational Tests. *Symmetry* **2016**, *8*, 111. [CrossRef]

37. Abbott, B.P.; Abbott, R.; Abbott, T.D.; Acernese, F.; Ackley, K.; Adams, C.; Adams, T.; Addesso, P.; Adhikari, R.X.; Adya, V.B.; et al. Gravitational Waves and Gamma-Rays from a Binary Neutron Star Merger: GW170817 and GRB 170817A. *Astrophys. J.* **2017**, *848*, L13. [CrossRef]
38. Bourgoin, A.; Hees, A.; Bouquillon, S.; Poncin-Lafitte, C. Le; Francou, G.; Angonin, M.C. Testing Lorentz symmetry with Lunar Laser Ranging. *Phys. Rev. Lett.* **2016**, *117*, 241301. [CrossRef] [PubMed]
39. Mueller, H. Atom interferometry tests of the isotropy of post-Newtonian gravity. *Phys. Rev. Lett.* **2008**, *100*, 031101. [CrossRef] [PubMed]
40. Flowers, N.A.; Goodge, C.; Tasson, J.D. Superconducting-Gravimeter Tests of Local Lorentz Invariance. *Phys. Rev. Lett.* **2017**, *119*, 201101. [CrossRef] [PubMed]
41. Shao, C.G.; Chen, Y.F.; Sun, R.; Cao, L.S.; Zhou, M.K.; Hu, Z.K.; Yu, C.; Müller, H. Limits on Lorentz violation in gravity from worldwide superconducting gravimeters. *Phys. Rev. D* **2018**, *97*, 024019. [CrossRef]

Article

Is There Any Symmetry Left in Gravity Theories with Explicit Lorentz Violation?

Yuri Bonder * and Cristóbal Corral

Instituto de Ciencias Nucleares, Universidad Nacional Autónoma de México
Apartado Postal 70-543, Ciudad de México 04510, Mexico; cristobal.corral@correo.nucleares.unam.mx
* Correspondence: bonder@nucleares.unam.mx

Received: 15 August 2018; Accepted: 18 September 2018; Published: 25 September 2018

Abstract: It is well known that a theory with explicit Lorentz violation is not invariant under diffeomorphisms. On the other hand, for geometrical theories of gravity, there are alternative transformations, which can be best defined within the first-order formalism and that can be regarded as a set of improved diffeomorphisms. These symmetries are known as local translations, and among other features, they are Lorentz covariant off shell. It is thus interesting to study if theories with explicit Lorentz violation are invariant under local translations. In this work, an example of such a theory, known as the minimal gravity sector of the Standard Model Extension, is analyzed. Using a robust algorithm, it is shown that local translations are not a symmetry of the theory. It remains to be seen if local translations are spontaneously broken under spontaneous Lorentz violation, which are regarded as a more natural alternative when spacetime is dynamic.

Keywords: local Lorentz invariance; diffeomorphism invariance; local translations; first-order formalism

---

1. Introduction

Conventional theories of gravity, when geometrical, are invariant under diffeomorphisms (Diff) and local Lorentz transformations (LLT). In addition, such theories are invariant under the so-called local translations (LT), which can be regarded as improved Diff in the sense that they are fully Lorentz covariant, among other properties [1]. As expected, for theories that are invariant under LLT, on shell, invariance under Diff implies invariance under LT and vice versa [2,3]. Thus, in light of this result, it is interesting to study the logical relation between Diff and LT when LLT is explicitly broken. This is the main goal of this work. Clearly, symmetry is a very powerful tool that simplifies calculations and provides conceptual clarity; therefore, elucidating if there is a symmetry left in theories with explicit Lorentz violation can be extremely relevant.

This work is motivated by the idea that general relativity is a low energy limit of a more fundamental theory of gravity that incorporates the quantum principles [4]. Several interesting quantum gravity approaches have been developed. The leading candidates include string theory, loop quantum gravity, spin networks, noncommutative geometry, causal sets and causal dynamical triangulations (see the corresponding contributions in [5]). Notably, even though this theory is still unknown, research fields like quantum cosmology are making steady progress by appealing to effects inspired by quantum gravity [6,7].

On the other hand, there is an approach to quantum gravity where the priority is to make contact with experiments and which is known as quantum gravity phenomenology. Within quantum gravity phenomenology, it is customary to look for traces of Lorentz violation. The motivation to do this stems from the fact that within the most prominent approaches to quantum gravity, it has been argued that LLT may not be a fundamental symmetry [8–12], and thus, it may be possible to look for empirical traces of quantum gravity by searching for violations of LLT. Inspired by this possibility, a parametrization of Lorentz violation based on effective field theory has been developed [13,14], which is known as the Standard Model Extension (SME).

The SME action contains Lorentz-violating extensions to all sectors of conventional physics, including general relativity [15]; this latter sector is known as the gravity sector. Moreover, within the gravity sector, the part that produces second-order field equations for the metric is called the minimal gravity sector, and using a post-Newtonian expansion [16], it has been tested with several interesting experiments including atom interferometry [17], frame dragging [18], lunar ranging [19,20], pulsar timing [21,22] and planetary motion [23]. There are also tests in the context of cosmology [24]. Other experiments related to the SME are reported in [25].

It should be stressed that the dominant position in the SME community is that, in the presence of gravity, Lorentz violation must occur spontaneously. This is assumed to deal with the severe restrictions arising from the contracted Bianchi identity [15], i.e., the fact that the Einstein tensor is divergence free. This, in turn, is closely related to the invariance under Diff. In addition, some relations between Diff and LLT are known in the literature. For example, it has been shown that spontaneous violation of Diff implies spontaneous violation of LLT, and vice versa [26–28]. In these regards, some surprising results have already been uncovered by using LT. In particular, in the unimodular theory of gravity [29], explicit breaking of Diff produces a breakdown of LT, but contrary to the expectations, LLT is unaffected [30]. Thus, it is interesting to analyze the fate of the LT when LLT is spontaneously broken. Here, however, attention is restricted to explicit symmetry breaking in the minimal gravity sector of the SME. This assumption is adopted for simplicity, but also since the dynamics associated with spontaneous Lorentz violation may spoil the Cauchy initial value formulation [31]. Moreover, this setup allows one to study if torsion, which modifies the Bianchi identities, can relax the restrictions that have driven the SME community to consider spontaneous Lorentz violation. Other alternatives to deal with these restrictions include a Stückelberg-like mechanism [32] and the use of Finsler geometries [33].

2. Gauge Theories of Gravity and Local Translations

Local translations can be best defined in gauge theories of gravity [1,34,35] that consider two independent gravitational fields: the tetrad $e^a{}_\mu$ and the Lorentz connection $\omega^{ab}{}_\mu = -\omega^{ba}{}_\mu$. This setup includes the well-known Poincaré gauge theories ([3], Chapter 3). Here, spacetime indices are represented by Greek letters and tangent-space indices with Latin characters; the summation convention on repeated indices is understood. The four-dimensional spacetime metric $g_{\mu\nu}$ is related to the tetrad by $g_{\mu\nu} = \eta_{ab}e^a{}_\mu e^b{}_\nu$, where $\eta_{ab} = \mathrm{diag}\,(-,+,+,+)$. Note that η_{ab} and its inverse, η^{ab}, can be used to lower and raise tangent-space indices, and that the tetrad and its inverse, $E^\mu{}_a$, which is such that $e^a{}_\mu E^\mu{}_b = \delta^a_b$ and $e^a{}_\mu E^\nu{}_a = \delta^\mu_\nu$, can be used to map spacetime indices to tangent-space indices and vice versa. In addition, the Lorentz connection and the spacetime connection $\Gamma^\lambda{}_{\mu\nu}$ are related through the tetrad postulate:

$$\partial_\mu e^a{}_\nu + \omega^a{}_{b\mu}e^b{}_\nu = \Gamma^\lambda{}_{\mu\nu}e^a{}_\lambda. \tag{1}$$

The left-hand side of this equation can be written as $\mathscr{D}_\mu e^a{}_\nu$ where $\mathscr{D}_\mu$ is the covariant derivative with respect to the Lorentz connection. Notice that the covariant derivative used here differs from the operator widely used in SME papers (e.g., [15]), which is represented by D_μ, in that, when acting on a tensor, $\mathscr{D}_\mu$ does not add a connection term for each spacetime index. These two operators are discussed in [36] in a notation that does not coincide with the one used here.

Curvature and torsion can be derived from the tetrad and the Lorentz connection through Cartan's structure equations:

$$\frac{1}{2}R^{ab}{}_{\mu\nu} = \partial_{[\mu}\omega^{ab}{}_{\nu]} + \omega^a{}_{c[\mu}\omega^{cb}{}_{\nu]}, \tag{2}$$

$$\frac{1}{2}T^a{}_{\mu\nu} = \mathscr{D}_{[\mu}e^a{}_{\nu]}, \tag{3}$$

where the squared brackets denote antisymmetrization of the n indices enclosed (with a $1/n!$ factor). Clearly, $R^{ab}_{\ \ \mu\nu} = R^{[ab]}_{\ \ [\mu\nu]}$ and $T^a_{\ \mu\nu} = T^a_{\ [\mu\nu]}$, and from Equations (1) and (3), it can be verified that $T^a_{\ \mu\nu}/2 = \Gamma^\lambda_{\ [\mu\nu]}e^a_{\ \lambda}$. Furthermore, the Bianchi identities take the form:

$$\mathscr{D}_{[\mu}R^{ab}_{\ \ \nu\lambda]} - T^c_{\ [\mu\nu}R^{ab}_{\ \ \lambda]\rho}E^\rho_{\ c} = 0, \tag{4}$$

$$\mathscr{D}_{[\mu}T^a_{\ \nu\lambda]} - T^b_{\ [\mu\nu}T^a_{\ \lambda]\rho}E^\rho_{\ b} = R^a_{\ [\mu\nu\lambda]}. \tag{5}$$

As usual, (active) infinitesimal Diff are implemented by the Lie derivative. Since the tetrad and the Lorentz connection are one-forms, it reads:

$$\text{Diff} = \begin{cases} \delta_{\text{Diff}}(\rho)e^a_{\ \mu} &= \rho^\nu\partial_\nu e^a_{\ \mu} + \partial_\mu\rho^\nu e^a_{\ \nu}, \\ \delta_{\text{Diff}}(\rho)\omega^{ab}_{\ \ \mu} &= \rho^\nu\partial_\nu\omega^{ab}_{\ \ \mu} + \partial_\mu\rho^\nu\omega^{ab}_{\ \ \nu}. \end{cases} \tag{6}$$

Moreover, under LLT, the tetrad and Lorentz connection respectively transform as a vector and gauge connection, that is:

$$\text{LLT} = \begin{cases} \delta_{\text{LLT}}(\lambda)e^a_{\ \mu} &= -\lambda^a_{\ b}e^b_{\ \mu}, \\ \delta_{\text{LLT}}(\lambda)\omega^{ab}_{\ \ \mu} &= \mathscr{D}_\mu\lambda^{ab}. \end{cases} \tag{7}$$

The fact that Diff involves partial derivatives and not covariant derivatives under LLT already suggests that Diff are not Lorentz covariant. However, this observation also points to its cure: define a transformation replacing the partial derivative by a covariant derivative. This is the most conventional way to introduce the LT [1] and the result, for the case where the gauge group is LLT, is:

$$\text{LT} = \begin{cases} \delta_{\text{LT}}(\rho)e^a_{\ \mu} = \mathscr{D}_\mu\rho^a + \rho^\nu T^a_{\ \nu\mu}, \\ \delta_{\text{LT}}(\rho)\omega^{ab}_{\ \ \mu} = \rho^\nu R^{ab}_{\ \ \nu\mu}, \end{cases} \tag{8}$$

where $\rho^a = e^a_{\ \mu}\rho^\mu$. Remarkably, it is easy to verify that, acting on $e^a_{\ \mu}$ or $\omega^{ab}_{\ \ \mu}$,

$$\delta_{\text{Diff}}(\rho) = \delta_{\text{LT}}(\rho) + \delta_{\text{LLT}}(\tilde\lambda), \tag{9}$$

where:

$$\tilde\lambda^{ab} = \rho^\mu\omega^{ab}_{\ \ \mu}. \tag{10}$$

From this relation, it can be verified that Diff and LT are equivalent in theories that are invariant under LLT. What is more, if a theory is invariant under two of these symmetries, it has to be invariant under the third. Conversely, this suggests that, if a theory breaks a symmetry, as is the case of the SME, it should break at least one of the remaining symmetries.

Recall that general relativity is invariant under the two symmetry classes: LLT and Diff. The former acts locally on the tangent space, and it thus can be regarded as the gauge (or internal) symmetry of the theory, while the latter connects different spacetime points. Note that, from Equation (9), it seems that the LT act in both, the tangent space, through LLT, and the manifold, via the Diff. However, the fact that the gravitational potentials transform under LT as Lorentz tensors, which does not occur under Diff, can be used to argue that they only act on the tangent space.

Now, it is possible to construct theories where the gauge group is different from LLT. Perhaps the most popular examples of such theories are those where the gauge group is de Sitter or anti-de Sitter [36–39]. Fortunately, the LT definition can be generalized to theories invariant under arbitrary gauge transformations (GT). Thus, for the sake of generality, in the remainder of this section, the role of the LLT is played by a general GT. To obtain the generalized LT, one simply needs to replace the partial derivatives, in Equation (6), by covariant derivatives associated with the corresponding gauge group. Equivalently, they can be defined [40] as the difference of $\delta_{\text{Diff}}(\rho)$ and $\delta_{\text{GT}}(\tilde\lambda)$, for the corresponding

$\tilde{\lambda}^{ab}$, thus generalizing Equation (9). It is important to remark that the action of LT on the geometrical fields depends on the gauge symmetry, and for theories invariant under generic GT, it does not need to coincide with Equation (8).

Interestingly, an algorithm has been developed [30] that, starting from the action, allows one to verify if the theory is invariant under LT and GT, and if it does, it gives the corresponding transformations of the dynamical fields (see also [41]). What is more, in certain cases, the algorithm leads to the corresponding contracted Bianchi identities and the matter conservation laws. Since the algorithm is closely related to Nöther's theorem, as can be seen from Equations (16) and (17), it selects the fundamental symmetries of the theory. In fact, the algorithm's output contains the transformation laws of the dynamical fields under GT and LT; whether the theory is invariant under Diff can be derived from such transformations. In this sense the GT and LT are more fundamental than the Diff. The covariant derivative associated with GT is denoted by $\tilde{\mathscr{D}}_\mu$, and the situation in which GT is the Lorentz group, which leads to Equation (8), arises as a particular case. Furthermore, for simplicity, the algorithm is done in a four-dimensional spacetime that has no boundaries, and all dynamical fields besides the tetrad and Lorentz connection, which are denoted by Ψ, are assumed to be zero-forms in a nontrivial representation of the Lorentz group. Notice that Ψ may be Dirac spinors. Also observe that the fact that the formalism is based on the action, and not the Hamiltonian, allows one to disregard all issues related with spacetime foliations.

The basic steps of the algorithm are: (i) Consider an action principle:

$$S[e^a{}_\mu, \omega^{ab}{}_\mu, \Psi] = \int \mathrm{d}^4 x \, e \, \mathscr{L}(e^a{}_\mu, \omega^{ab}{}_\mu, \Psi), \tag{11}$$

where $\mathrm{d}^4 x \, e$ is the covariant four-volume element and $\mathscr{L}$ is an arbitrary Lagrangian. (ii) Perform an arbitrary variation of the action with respect to the dynamical fields:

$$\delta S = \int \mathrm{d}^4 x \, e \left(\delta e^a{}_\mu F^\mu{}_a + \delta \omega^{ab}{}_\mu F^\mu{}_{ab} + \delta \Psi F \right), \tag{12}$$

which implicitly defines $F^\mu{}_a$, $F^\mu{}_{ab}$ and F. Note that $F^\mu{}_a$ includes the variation of the volume element, namely $F^\mu{}_a = (1/e)\delta(e\mathscr{L})/\delta e^a_\mu$, and that these objects vanish on shell. Then, (iii) apply a covariant derivative $\tilde{\mathscr{D}}_\mu$ to $F^\mu{}_a$, $F^\mu{}_{ab}$ and F. Step (iv): Verify if the resulting expressions can be written as linear combinations of $F^\mu{}_a$, $F^\mu{}_{ab}$ and F, where the coefficients can be spacetime functions. If possible, it reads:

$$\tilde{\mathscr{D}}_\mu F^\mu{}_a = \alpha_\mu F^\mu{}_a + \beta^b{}_\mu F^\mu{}_{ab} + \gamma_a F, \tag{13}$$

$$\tilde{\mathscr{D}}_\mu F^\mu{}_{ab} = \pi_{\mu[a} F^\mu{}_{b]} + \theta_\mu F^\mu{}_{ab} + \mu_{[ab]} F, \tag{14}$$

$$\tilde{\mathscr{D}}_\mu F = 0, \tag{15}$$

where the fact that F is a four-form is used. Note that the precise form of the coefficients in Equations (13) and (14) depends on the theory and, in particular, on the gauge group. Step (v): Multiply Equations (13) and (14), respectively, with gauge parameters $\rho^a(x)$ and $\xi^{ab}(x) = -\xi^{ba}(x)$. Finally, (vi) integrate over spacetime using the appropriate four-volume element. At this stage, one must use that $\tilde{\mathscr{D}}_\mu$ can be replaced by ∂_μ when acting on a scalar under the gauge group. Furthermore, the Leibniz rule, Gauss's theorem and the assumption that there are no spacetime boundaries are utilized to take the equations to the form:

$$0 = \int d^4x\, \partial_\mu \left(e\, F^\mu{}_a \rho^a \right)$$

$$= \int d^4x\, e\, \big[\underbrace{\left(\rho^a \partial_\mu \ln e + \mathscr{D}_\mu \rho^a + \alpha_\mu \rho^a \right)}_{\delta_{\mathrm{LT}}(\rho)e^a{}_\mu} F^\mu{}_a + \underbrace{\rho^{[a}\beta^{b]}{}_\mu}_{\delta_{\mathrm{LT}}(\rho)\omega^{ab}{}_\mu} F^\mu{}_{ab} + \underbrace{\gamma_a \rho^a}_{\delta_{\mathrm{LT}}(\rho)\Psi} F \big], \tag{16}$$

$$0 = \int d^4x\, \partial_\mu \left(e\, F^\mu{}_{ab} \zeta^{ab} \right)$$

$$= \int d^4x\, e\, \big[\underbrace{-\pi_{\mu b}\zeta^{ab}}_{\delta_{\mathrm{GT}}(\zeta)e^a{}_\mu} F^\mu{}_a + \underbrace{\left(\zeta^{ab}\partial_\mu \ln e + \bar{\mathscr{D}}_\mu \zeta^{ab} + \theta_\mu \zeta^{ab} \right)}_{\delta_{\mathrm{GT}}(\zeta)\omega^{ab}{}_\mu} F^\mu{}_{ab} + \underbrace{\mu_{ab}\zeta^{ab}}_{\delta_{\mathrm{GT}}(\zeta)\Psi} F \big]. \tag{17}$$

Then, comparing with the action variation (12), it is possible to read off the field transformations under LT and GT. On the other hand, if it is impossible to write the covariant derivatives of $F^\mu{}_a$ and $F^\mu{}_{ab}$ as in Equations (13) and (14), then one can identify which symmetries are broken and which terms are responsible for such breakdowns. In the next section, this algorithm is applied to the minimal gravity sector of the SME.

3. Explicit Lorentz Violation in the Gravity Sector

A theory with explicit Lorentz violation is one in which identical experiments done in different inertial frames can produce different results. It is thus easy to imagine experiments designed to look for such violations. In fact, the Earth's rotation (translation) gives a natural family of instantaneous inertial frames in which Lorentz violations can manifest themselves as signals with a daily (yearly) period. Moreover, there are concrete models that incorporate Lorentz violation to account for astrophysical puzzles like the presence of cosmic rays above the GZK cutoff [42]. Here, attention is restricted to a sector of the SME, which should be regarded as a generic parametrization of Lorentz violation.

The action of the minimal gravity sector of the SME, in the first-order formalism, takes the form $S = S_g + S_m$, where the Lorentz-violating gravitational part is:

$$S_g[e^a{}_\mu, \omega^{ab}{}_\mu] = \frac{1}{2\kappa} \int d^4x\, e\, E^\mu{}_a E^\nu{}_b \left(R^{ab}{}_{\mu\nu} + k_{cd}{}^{ab} R^{cd}{}_{\mu\nu} \right). \tag{18}$$

Here, $\kappa = 8\pi G_N$ is the gravitational coupling constant, and $e = \det e^a{}_\mu$. The first term of this action is the Einstein–Hilbert action, and $k_{cd}{}^{ab}$ is a non-dynamical zero-form parametrizing possible Lorentz violations and whose components are known as the SME coefficients. The SME coefficients are typically expressed in terms of spacetime indices, and this can be achieved by using the tetrad and its inverse to translate the Latin indices in $k_{cd}{}^{ab}$ to spacetime indices. From the index symmetries of $R^{ab}{}_{\mu\nu}$, it is clear that $k_{cd}{}^{ab} = k_{[cd]}{}^{[ab]}$. However, from Equation (5), it can be seen that, in the presence of torsion, it is not necessary to assume that $k_{a[bcd]} = 0$. Thus, there are 36 independent components of $k_{cd}{}^{ab}$; this should be compared with the 20 independent components that are present when torsion vanishes. Note that the SME coefficients sensitive to torsion are studied in [15]. For simplicity, even though it has been shown that there is a York–Gibbons–Hawking term for the minimal gravity SME sector [43], spacetime boundaries are not considered. Furthermore, the SME coefficients are assumed to be extremely small in any relevant reference frame; thus, they should not damage the Cauchy initial value formulation of general relativity [44].

The matter action is:

$$S_m = \int d^4x\, e\, \mathscr{L}_m[e^a{}_\mu, \omega^{ab}{}_\mu, \Psi], \tag{19}$$

where Ψ denotes the matter fields, which are taken to be zero-forms in a nontrivial representation of the Lorentz group (recall that Ψ includes spinors). As is customary, the energy-momentum and spin densities are respectively defined by:

$$\tau^{\mu}{}_{a} = \frac{1}{e}\frac{\delta(e\mathcal{L}_m)}{\delta e^{a}{}_{\mu}}, \tag{20}$$

$$\sigma^{\mu}{}_{ab} = 2\frac{\delta\mathcal{L}_m}{\delta\omega^{ab}{}_{\mu}}. \tag{21}$$

Furthermore, since the gravity action does not depend on Ψ, it is possible to define, using the notation of the previous section, $F = \delta\mathcal{L}_m/\delta\Psi$. The main assumption on S_m is that it is invariant under Diff and LLT. This leads to two off-shell conservation laws:

$$(\mathcal{D}_{\mu} + T_{\mu})\tau^{\mu}{}_{a} = T^{b}{}_{\mu\nu}E^{\mu}{}_{a}\tau^{\nu}{}_{b} + \frac{1}{2}R^{bc}{}_{\mu\nu}E^{\mu}{}_{a}\sigma^{\nu}{}_{bc} - FE^{\mu}{}_{a}\partial_{\mu}\Psi$$

$$- \frac{1}{2}E^{\mu}{}_{a}\omega^{bc}{}_{\mu}[(\mathcal{D}_{\nu} + T_{\nu})\sigma^{\nu}{}_{bc} - 2g_{\nu\rho}E^{\nu}{}_{b}\tau^{\rho}{}_{c}], \tag{22}$$

$$(\mathcal{D}_{\mu} + T_{\mu})\sigma^{\mu}{}_{ab} = 2g_{\mu\nu}E^{\mu}{}_{[a}\tau^{\nu}{}_{b]} + FJ_{ab}\Psi, \tag{23}$$

where $T_{\mu} = T^{a}{}_{\mu\nu}E^{\nu}{}_{a}$ and J_{ab} are the generators of LLT associated with Ψ, i.e., $\delta_{\text{LLT}}(\lambda)\Psi = -\lambda^{ab}J_{ab}\Psi/2$. Notice that the presence of a T_{μ} term next to every $\mathcal{D}_{\mu}$ is closely related to the presence of $\partial_{\mu}\ln e$ in the transformations that are read off from Equations (16) and (17).

Equation (22) is the generalization of the energy-momentum conservation law. In this equation, the issues associated with Diff invariance, which are mentioned in Section 2, become evident: there is a partial derivative of Ψ and the Lorentz connection appears explicitly and these two terms are not covariant under LLT. Of course, this is not an issue when Equation (23) is valid because, when these two equations are put together, the term with the Lorentz connection gets replaced by a term with J_{ab} that combines with the partial derivative to transform covariantly. In addition, since these terms are multiplied by F, they vanish on-shell. Of course, the LT are precisely built in such a way that the problematic terms do not even arise.

An arbitrary variation of the total action is given by Equation (12) with:

$$F^{\mu}{}_{a} = -\tau^{\mu}{}_{a} - \frac{1}{\kappa}\left[G^{\mu}{}_{a} - \frac{1}{2}k_{bc}{}^{de}R^{bc}{}_{\rho\sigma}\left(2E^{\mu}{}_{e}E^{\rho}{}_{a}E^{\sigma}{}_{d} + E^{\mu}{}_{a}E^{\rho}{}_{d}E^{\sigma}{}_{e}\right)\right], \tag{24}$$

$$F^{\mu}{}_{ab} = -\frac{1}{2}\sigma^{\mu}{}_{ab} + \frac{1}{2\kappa}\left[\left(T^{c}{}_{\rho\sigma} + 2e^{c}{}_{\rho}T_{\sigma}\right)E^{\rho}{}_{[a}E^{\sigma}{}_{b]}E^{\mu}{}_{c} + k_{ab}{}^{cd}T^{e}{}_{\rho\sigma}E^{\rho}{}_{c}E^{\sigma}{}_{d}E^{\mu}{}_{e}\right.$$

$$\left. -2E^{\nu}{}_{c}E^{\mu}{}_{d}(\mathcal{D}_{\nu} + T_{\nu})k_{ab}{}^{cd}\right], \tag{25}$$

where $G^{a}{}_{\mu} = R^{bc}{}_{\rho\sigma}(\delta^{a}_{b}\delta^{\rho}_{\mu}E^{\sigma}{}_{c} - e^{a}{}_{\mu}E^{\rho}{}_{b}E^{\sigma}{}_{c}/2)$ is the Einstein tensor in the presence of torsion. Thus, on-shell, Equation (24) is the generalization of the Einstein equation, which is not necessarily symmetric, and Equation (25) plays the role of the so-called Cartan equation.

The key step in the algorithm presented above is to take the covariant derivative of Equations (24) and (25). The results are the contracted Bianchi identities, which, using Equations (22) and (23), can be cast into the form:

$$(\mathcal{D}_{\mu} + T_{\mu})F^{\mu}{}_{a} = T^{b}{}_{\mu\nu}E^{\mu}{}_{a}F^{\nu}{}_{b} + R^{bc}{}_{\mu\nu}E^{\mu}{}_{a}F^{\nu}{}_{bc} - E^{\mu}{}_{a}\left(\partial_{\mu}\Psi + \frac{1}{2}\omega^{ab}{}_{\mu}J_{ab}\Psi\right)F$$

$$- \frac{1}{2\kappa}R^{bc}{}_{\rho\sigma}E^{\mu}{}_{a}E^{\rho}{}_{d}E^{\sigma}{}_{e}\mathcal{D}_{\mu}k_{bc}{}^{de}, \tag{26}$$

$$(\mathcal{D}_{\mu} + T_{\mu})F^{\mu}{}_{ab} = 2g_{\mu\nu}E^{\mu}{}_{[a}F^{\nu}{}_{b]} + \frac{1}{\kappa}\left(R^{cd}{}_{\mu\nu}k_{cd[a}{}^{e}E^{\mu}{}_{b]}E^{\nu}{}_{e} - R^{c}{}_{[a|\mu\nu|}k_{b]c}{}^{de}E^{\mu}{}_{d}E^{\nu}{}_{e}\right). \tag{27}$$

Clearly, the term with $\mathcal{D}_{\mu}k_{ab}{}^{cd}$ in Equation (26) breaks LT invariance since it cannot be written as a linear combination of $F^{\mu}{}_{a}$, $F^{\mu}{}_{ab}$ or F. Analogously, in Equation (27), the terms in the parenthesis break LLT. This answers the central question of this paper: the theory is generically not invariant under any of the symmetries considered. Still, by making the last term in Equation (26) equal to zero while letting the parenthesis in Equation (27) be nonzero, one can break LLT while the theory is invariant under LT. Clearly, it is also possible to have invariance under LLT while LT is broken. However, tuning the values of the SME coefficients goes against the SME philosophy of keeping them arbitrary until they are constrained by experiments.

Remarkably, from Equation (25), it can be realized that $k_{ab}{}^{cd}$ acts as a torsion source, as it occurs in theories with nonminimal scalar couplings (see [45] and the references therein). Thus, even in vacuum, the presence of $k_{ab}{}^{cd}$ generates torsion. This is very unusual since, in most theories, vacuum torsion vanishes. Moreover, this torsion field could be probed with spinors. In fact, for a Dirac spinor ψ, the equation of motion for ψ takes the form [46]:

$$iE^{\mu}{}_{a}\gamma^{a}\left(\partial_{\mu}\psi + \frac{i}{4}\tilde{\omega}^{bc}{}_{\mu}\sigma_{bc}\psi\right) + E^{\mu}{}_{a}A_{\mu}\gamma_{5}\gamma^{a}\psi = 0, \tag{28}$$

where γ^{a} are the Dirac matrices satisfying Clifford's algebra $\{\gamma_{a}, \gamma_{b}\} = -2\eta_{ab}$, $\sigma_{ab} = i\gamma_{[a}\gamma_{b]}$ and $\gamma_{5} = i\gamma^{0}\gamma^{1}\gamma^{2}\gamma^{3}$. In addition, $\tilde{\omega}^{ab}{}_{\mu}$ is the torsion-free spin connection (i.e., it satisfies the torsion-free tetrad postulate) and $A^{\mu} = \epsilon^{\rho\sigma\nu\mu}T_{\rho\sigma\nu}/6$, with $\epsilon_{\mu\nu\rho\sigma}$ the Levi–Civita tensor. Notably, bounds on vacuum torsion like those discussed in [46] could be motivated by the presence of torsion in the minimal gravity sector of the SME. On the other hand, if torsion is assumed to be nonzero, one can use the very stringent bounds on the matter sector of the SME (see [25]) to put limits on $k_{ab}{}^{cd}$; this is similar to what can be done using field redefinitions in the matter-gravity SME sector [47]. Furthermore, the conventional SME phenomenology is recovered when torsion and spin density are both set to zero. Notice that the former can be rigorously turned off with a Lagrange multiplier [48]. However, even in this case, it is expected that the LT will not be a symmetry of the theory since the structure of Equations (26) and (27) does not change by the presence of this Lagrange multiplier.

Finally, as is mentioned in the Introduction, in the SME community, Equation (26) is seen as a strong restriction linking matter, geometry and k_{abcd}. However, when torsion and spin density are considered, this condition is relaxed in the sense that it involves more degrees of freedom of both the matter and the geometry. This may be a valuable alternative in addition to spontaneous Lorentz violation.

4. Conclusions

In this work, the so-called local translations are studied in a gravity theory with explicit Lorentz violation, which is introduced by a non-dynamical zero-form $k_{ab}{}^{cd}$. It was known that the theory is not diffeomorphism invariant, and in this work, it is shown that the theory is also not invariant under translational invariance. This is interesting since the local translations can be regarded as improved diffeomorphisms in the sense that they are fully covariant under local Lorentz transformations, thus having the potential to be unaffected by Lorentz violation. Another interesting aspect that becomes evident is that the minimal gravity sector of the SME, in the presence of torsion, has additional coefficients, which generate vacuum torsion.

There are interesting issues that could be addressed in future contributions. For example, what happens with the local translations in theories with spontaneous Lorentz violation? In this case, $k_{ab}{}^{cd}$ would be dynamical, and it thus transforms under all symmetries. Furthermore, even if the local translations are indeed spontaneously broken, there could still be advantages of using them when there is spontaneous Lorentz violation. The ultimate role of these local translations needs to be carefully analyzed but, if some traces of them remain, they could play very important roles.

Author Contributions: The authors contributed equally.

Funding: This research was funded by UNAM-DGAPA-PAPIIT Grant No. RA101818 and a UNAM-DGAPA postdoctoral fellowship.

Acknowledgments: The authors acknowledge getting valuable input from D. González, A. Kostelecký and M. Montesinos .

Conflicts of Interest: The authors declare no conflict of interest.

References

1. Hehl, F.W.; McCrea, J.D.; Mielke, E.W.; Ne'eman, Y. Metric affine gauge theory of gravity: Field equations, Noether identities, world spinors, and breaking of dilation invariance. *Phys. Rept.* **1995**, *258*, 1–171. [CrossRef]
2. Hehl, F.W.; Von Der Heyde, P.; Kerlick, G.D.; Nester, J.M. General relativity with spin and torsion: Foundations and prospects. *Rev. Mod. Phys.* **1976**, *48*, 393. [CrossRef]
3. Blagojevic, M. *Gravitation and Gauge Symmetries*; CRC Press: Boca Raton, FL, USA, 2002.
4. Burgess, C.P. Quantum Gravity in Everyday Life: General Relativity as an Effective Field Theory. *Liv. Rev. Relativ* **2004**, *7*. [CrossRef]
5. Oriti, D. *Approaches to Quantum Gravity: Toward a New Understanding of Space, Time and Matter*; Cambridge University Press: Cambridge, UK, 2009.
6. Kiefer, C.; Krämer, M. Quantum Gravitational Contributions to the Cosmic Microwave Background Anisotropy Spectrum. *Phys. Rev. Lett.* **2012**, *108*, 021301. [CrossRef]
7. Capozziello, S.; Luongo, O. Entanglement inside the cosmological apparent horizon. *Phys. Lett. A* **2014**, *378*, 2058–2062. [CrossRef]
8. Kostelecký, V.A.; Samuel, S. Spontaneous breaking of Lorentz symmetry in string theory. *Phys. Rev. D* **1989**, *39*. [CrossRef]
9. Kostelecký, V.A.; Potting, R. CPT and strings. *Nucl. Phys. B* **1991**, *359*, 545–570. [CrossRef]
10. Gambini, R.; Pullin, J. Nonstandard optics from quantum space-time. *Phys. Rev. D* **1999**, *59*. PhysRevD.59.124021. [CrossRef]
11. Alfaro, J.; Morales-Técotl, H.A.; Urrutia, L.F. Quantum Gravity Corrections to Neutrino Propagation. *Phys. Rev. Lett.* **2000**, *84*, 2318–2321. [CrossRef]
12. Carroll, S.M.; Harvey, J.A.; Kostelecký, V.A.; Lane, C.D.; Okamoto, T. Noncommutative Field Theory and Lorentz Violation. *Phys. Rev. Lett.* **2001**, *87*. [CrossRef]
13. Colladay, D.; Kostelecký, V.A. CPT violation and the standard model. *Phys. Rev. D* **1997**, *55*, 6760–6774. [CrossRef]
14. Colladay, D.; Kostelecký, V.A. Lorentz-violating extension of the standard model. *Phys. Rev. D* **1998**, *58*. [CrossRef]
15. Kostelecký, V.A. Gravity, Lorentz violation, and the standard model. *Phys. Rev. D* **2004**, *69*. [CrossRef]
16. Bailey, Q.G.; Kostelecký, V.A. Signals for Lorentz violation in post-Newtonian gravity. *Phys. Rev. D* **2006**, *74*. [CrossRef]
17. Müller, H.; Chiow, S.W.; Herrmann, S.; Chu, S.; Chung, K.Y. Atom-Interferometry Tests of the Isotropy of Post-Newtonian Gravity. *Phys. Rev. Lett.* **2008**, *100*. [CrossRef]
18. Bailey, Q.G.; Everett, R.D.; Overduin, J.M. Limits on violations of Lorentz symmetry from Gravity Probe B. *Phys. Rev. D* **2013**, *88*. [CrossRef]
19. Battat, J.B.R.; Chandler, J.F.; Stubbs, C.W. Testing for Lorentz Violation: Constraints on Standard-Model-Extension Parameters via Lunar Laser Ranging. *Phys. Rev. Lett.* **2007**, *99*. [CrossRef]
20. Bourgoin, A.; Hees, A.; Bouquillon, S.; Le Poncin-Lafitte, C.; Francou, G.; Angonin, M.C. Testing Lorentz Symmetry with Lunar Laser Ranging. *Phys. Rev. Lett.* **2016**, *117*. [CrossRef]
21. Shao, L. Tests of Local Lorentz Invariance Violation of Gravity in the Standard Model Extension with Pulsars. *Phys. Rev. Lett.* **2014**, *112*. [CrossRef]
22. Shao, L. New pulsar limit on local Lorentz invariance violation of gravity in the standard-model extension. *Phys. Rev. D* **2014**, *90*. [CrossRef]
23. Hees, A.; Bailey, Q.G.; Le Poncin-Lafitte, C.; Bourgoin, A.; Rivoldini, A.; Lamine, B.; Meynadier, F.; Guerlin, C.; Wolf, P. Testing Lorentz symmetry with planetary orbital dynamics. *Phys. Rev. D* **2015**, *92*. [CrossRef]
24. Bonder, Y.; León, G. Inflation as an amplifier: The case of Lorentz violation. *Phys. Rev. D* **2017**, *96*. [CrossRef]
25. Kostelecký, V.A.; Russell, N. Data tables for Lorentz and CPT violation. *Rev. Mod. Phys.* **2011**, *83*. [CrossRef]
26. Bluhm, R.; Kostelecký, V.A. Spontaneous Lorentz violation, Nambu–Goldstone modes, and gravity. *Phys. Rev. D* **2005**, *71*. [CrossRef]

27. Bluhm, R.; Fung, S.H.; Kostelecký, V.A. Spontaneous Lorentz and diffeomorphism violation, massive modes, and gravity. *Phys. Rev. D* **2008**, *77*. [CrossRef]
28. Kostelecký, V.A.; Mewes, M. Lorentz and diffeomorphism violations in linearized gravity. *Phys. Lett. B* **2018**, *779*, 136–142. [CrossRef]
29. Bonder, Y.; Corral, C. Unimodular Einstein–Cartan gravity: Dynamics and conservation laws. *Phys. Rev. D* **2018**, *97*. [CrossRef]
30. Corral, C.; Bonder, Y. Local translations in modified gravity theories. *arXiv*, **2018**, arXiv:1808.01497.
31. Bonder, Y.; Escobar, C.A. Dynamical ambiguities in models with spontaneous Lorentz violation. *Phys. Rev.D* **2016**, *93*. [CrossRef]
32. Bluhm, R. Gravity Theories with Background Fields and Spacetime Symmetry Breaking. *Symmetry* **2017**, *9*, 230. [CrossRef]
33. Kostelecký, V.A. Riemann–Finsler geometry and Lorentz-violating kinematics. *Phys. Lett. B* **2011**, *701*. [CrossRef]
34. Obukhov, Y.N. Poincare gauge gravity: Selected topics. *Int. J. Geom. Meth. Mod. Phys.* **2006**, *3*, 95–137. [CrossRef]
35. Blagojević, M.; Hehl, F.W. *Gauge Theories of Gravitation*; World Scientific Publishing: Singapore, 2013.
36. Hassaine, M.; Zanelli, J. *Chern–Simons (Super)Gravity*; World Scientific Publishing: Singapore, 2016.
37. MacDowell, S.W.; Mansouri, F. Unified Geometric Theory of Gravity and Supergravity. *Phys. Rev. Lett.* **1977**, *38*, 739. Erratum: *Phys. Rev. Lett.* **1977**, *38*, 1376. [CrossRef]
38. Stelle, K.S.; West, P.C. De Sitter gauge invariance and the geometry of the Einstein-Cartan theory. *J. Phys. A* **1979**, *12*, L205. [CrossRef]
39. Troncoso, R.; Zanelli, J. Higher dimensional gravity, propagating torsion and AdS gauge invariance. *Class. Quantum Grav.* **2000**, *17*, 4451–4466. [CrossRef]
40. Obukhov, Yu.N.; Rubilar, G.F. Invariant conserved currents in gravity theories: Diffeomorphisms and local gauge symmetries. *Phys. Rev. D* **2007**, *76*. [CrossRef]
41. Montesinos, M.; González, D.; Celada, M.; Díaz, B. Reformulation of the symmetries of first-order general relativity. *Class. Quantum Grav.* **2017**, *34*. [CrossRef]
42. Coleman, S.; Glashow, S.L. High-energy tests of Lorentz invariance. *Phys. Rev. D* **1999**, *59*. PhysRevD.59.116008. [CrossRef]
43. Bonder, Y. Lorentz violation in the gravity sector: The t puzzle. *Phys. Rev. D* **2015**, *91*. PhysRevD.91.125002. [CrossRef]
44. Ringström, H. *On the Topology and Future Stability of the Universe*; Oxford Science Publications: Oxford, UK, 2013.
45. Barrientos, J.; Cordonier-Tello, F.; Izaurieta. F.; Medina, P.; Narbona, D.; Rodríguez, E.; Valdivia, O. Nonminimal couplings, gravitational waves, and torsion in Horndeski's theory. *Phys. Rev. D* **2017**, *96*. [CrossRef]
46. Kostelecký, V.A.; Russell, N.; Tasson, J.D. Constraints on Torsion from Bounds on Lorentz Violation. *Phys. Rev. Lett.* **2008**, *100*. [CrossRef]
47. Bonder, Y. Lorentz violation in a uniform Newtonian gravitational field. *Phys. Rev. D* **2013**, *88*. [CrossRef]
48. del Pino, S.; Giribet, G.; Toloza, A.; Zanelli. J. From Lorentz-Chern–Simons to Massive Gravity in 2 + 1 dimensions. *JHEP* **2015**, *2015*. . [CrossRef]

Article

(Gravitational) Vacuum Cherenkov Radiation

Marco Schreck

Departamento de Física, Universidade Federal do Maranhão (UFMA), Campus Universitário do Bacanga, São Luís 65080-805, Brazil; marco.schreck@gmx.de; Tel.: +55-(98)-3272-8293

Received: 1 September 2018; Accepted: 15 September 2018; Published: 21 September 2018

Abstract: This work reviews our current understanding of Cherenkov-type processes in vacuum that may occur due to a possible violation of Lorentz invariance. The description of Lorentz violation is based on the Standard Model Extension (SME). To get an overview as general as possible, the most important findings for vacuum Cherenkov radiation in Minkowski spacetime are discussed. After doing so, special emphasis is put on gravitational Cherenkov radiation. For a better understanding, the essential properties of the gravitational SME are recalled in this context. The common grounds and differences of vacuum Cherenkov radiation in Minkowski spacetime and in the gravity sector are emphasized.

Keywords: Lorentz and CPT violation; standard model extension; vacuum Cherenkov radiation; modified gravity; experimental tests of gravity; gravitational waves

---

1. Introduction

Cherenkov radiation emerges when a massive, electrically charged particle travels through an optical medium with a velocity v that is larger than the phase velocity c_{med} of light in that medium. In fact, the characteristic, blueish glow that can be observed in the cooling water of nuclear reactors corresponds to this particular type of radiation. The effect was discovered by the Russians Cherenkov and Vavilov in 1934 [1]. Just about three years later, a theoretical explanation of their observation was provided by Frank and Tamm in the context of classical electromagnetism [2]. Cherenkov, Frank, and Tamm shared the Nobel prize of 1958. The fact that Vavilov was excluded is the likely explanation for why his name is usually omitted when referring to this phenomenon. The standard book reference on Cherenkov radiation is [3], which delivers explanations for various aspects of the effect. Although the book was first published several decades ago, most of the material covered is still valid and useful.

A microscopic explanation for Cherenkov radiation is as follows. As long as the charged particle travels with a speed $v < c_{\mathrm{med}}$, the few atoms or molecules polarized in the vicinity of the trajectory emit their wave trains out of phase leading to destructive interference. A particle propagating with a high velocity $v \geq c_{\mathrm{med}}$ enables the atoms or molecules of the medium to emit their wave trains in phase. Hence, wave trains interfere constructively producing coherent radiation far away from the source, which corresponds to the radiation observed. Constructive interference occurs on the surface of a Mach cone that bears the name Cherenkov cone. The radiation is emitted in directions perpendicular to the cone. These directions enclose an angle θ_c with the trajectory of the particle where θ_c is the Cherenkov angle. The latter is usually small, which means that Cherenkov radiation is emitted almost collinearly with the propagating particle. On the contrary, the Cherenkov cone is wide open.

Violations of Lorentz invariance are predicted by certain approaches to physics at the Planck scale [4,5]. An effective description of Lorentz violation, which is supposed to be valid for energies much smaller than the Planck energy, is known as the Standard Model Extension (SME) [6,7]. The SME parameterizes Lorentz and CPT-violation in terms of background fields that are properly contracted with field operators. A background field remains fixed under Lorentz transformations of matter fields and it decomposes into controlling coefficients that describe the magnitude of Lorentz violation.

The existence of a Lorentz-violating background field for a photon can also be interpreted as a vacuum with a nontrivial refractive index. The latter can depend on photon energy, direction, and polarization leading to the effects of dispersion, anisotropy, and birefringence, respectively. It is expected that a nontrivial vacuum may be accompanied by the presence of some kind of Cherenkov-type effect. As this effect occurs in vacuo, it is reasonable to refer to it as vacuum Cherenkov radiation.

In pre-relativity times, Sommerfeld was probably the first one to mention the concept of vacuum Cherenkov radiation [8–10]. Many decades later, Beall referred to it in the context of modified gravitational couplings [11]. Coleman and Glashow presumably performed the first modern investigation in [12]. The purpose of the current article is to provide a review on the research carried out on vacuum Cherenkov radiation since the point of time around the advent of the SME. For the past 20 years, a large number of articles has appeared shedding light on particular aspects of this interesting process [13–30]. Studies have been carried out within classical field theory [13,14,16,17,22,23,30] and quantum theory for spinless particles [18] as well as elementary fermions [15,18–20,27,28]. The latter references deal with effective theories in the sense that the authors consider pointlike approximations of extended particles such as mesons, hadrons, and even atomic nuclei. The substructure of hadrons was taken into account in some papers where sophisticated analyses were carried out based on parton distribution functions of the proton and neutron [24,29]. It was also discovered that Cherenkov-type processes are not limited to the emission of photons [21,26,29]. For example, as neutrinos are not electrically charged, they cannot emit photons. However, they can still lose energy in the presence of Lorentz violation by radiating Z bosons [21]. Last but not least, vacuum Cherenkov radiation can occur in the context of a modified gravity [25]. Lorentz violation in the pure-gravity sector modifies the dispersion relation and, therefore, the phase speed of gravitons. In a scenario where particles can travel faster than gravitons, they may lose energy by graviton emission. Special emphasis will be put on the SME gravity sector and the Cherenkov process associated with it.

2. Fundamentals of the Process

To understand the physics of vacuum Cherenkov radiation qualitatively, it is reasonable to study the structure of the nullcone and mass shell in momentum space. This nullcone is a surface plot of the photon dispersion law $\omega = \omega(\mathbf{p})$ as a function of the momentum $\mathbf{p}$, whereas the mass shell is an analog surface plot of the massive dispersion law $E = E(\mathbf{p})$. For the studies that we are interested in, it is sufficient to consider one-dimensional slices of these surfaces as shown in Figures 1 and 2. Vacuum Cherenkov radiation does not occur in the standard case simply due to the fact that the nullcone approaches the mass shell asymptotically for large momenta. Hence, it is not possible to shift one of the straight lines of the sliced nullcone such that it intersects the mass shell in two different points.

Vacuum Cherenkov radiation can emerge either for Lorentz violation in the sector of force carriers or for Lorentz violation in the matter sector, i.e., when either the nullcone or the mass shell (or both) are deformed. We will consider two examples of the first scenario. The first example is a quantum electrodynamics with an isotropic modification of the photon sector that is characterized by a controlling coefficient $\widetilde{\kappa}_{\mathrm{tr}} > 0$. The modified nullcone structure is shown in Figure 1a. In this case, the opening angle of the nullcone is larger in comparison to the standard theory, which means that photons travel with a phase velocity $v_{\mathrm{ph}} \equiv \omega(\mathbf{p})/|\mathbf{p}| < 1$. Now, it is possible without any problems to draw lines parallel to the nullcone such that they intersect the mass shell at two distinct points. Therefore, the massive particle will lose energy by photon emission where the recoil due to the photon emitted can be large enough to reverse the direction of the particle momentum. An analog process with a final-state photon of lower energy can take place subsequently, as we can again construct a line parallel to the nullcone having two intersections with the mass shell. In the second emission process, the recoil of the emitted photon is not sufficient anymore to reverse the momentum direction of the massive particle. After the two photon emissions, the mass shell can no longer be intersected twice with lines parallel to the nullcone, i.e., the massive particle ceases to radiate photons. The second example in the scenario of modified photons is shown in Figure 1b. It is based on an anisotropic Maxwell–Chern–Simons (MCS) term in the photon sector with dimensionful coefficient $k^3_{AF} = 5m_\psi$, where m_ψ is the fermion mass. Here, the nullcone is totally deformed due to dispersion. Parts of the deformed nullcone can be shifted to connect pairs of points on the standard mass shell such that the lower points are identified with the minimum of the deformed nullcone. For both photon emissions shown, the recoil is large enough to reverse the momentum direction of the radiating particle. The construction illustrated in Figure 2b can be perpetuated such that photons of ever increasing energy are emitted. This shows that vacuum Cherenkov radiation need not necessarily be a threshold process.

Vacuum Cherenkov radiation can also occur when the fermion sector is altered. The first example deals with a spin-degenerate modification of Dirac theory where the curvature of the modified mass shell is larger compared to the standard case, cf. Figure 2a. Two points on the mass shell can then be connected with lines parallel to the standard nullcone. Several photons may be emitted subsequently where the fermion momentum direction reverses by each emission. An alternative interesting possibility is that of a spin-nondegenerate modification of Dirac fermions shown in Figure 2b. In such a case, the modified mass shell splits into two distinct branches dependent on the spin direction of the fermion with respect to the quantization axis. The process qualitatively differs from that illustrated in Figure 2a, as one point sitting on a certain branch can be connected with a second point on the other branch by a line parallel to the standard nullcone. Photon emission takes place once and ceases when the second branch is reached. Furthermore, the direction of the fermion momentum does not reverse anymore, as a helicity change of the fermion is already caused by a spin flip due to the change of branch.

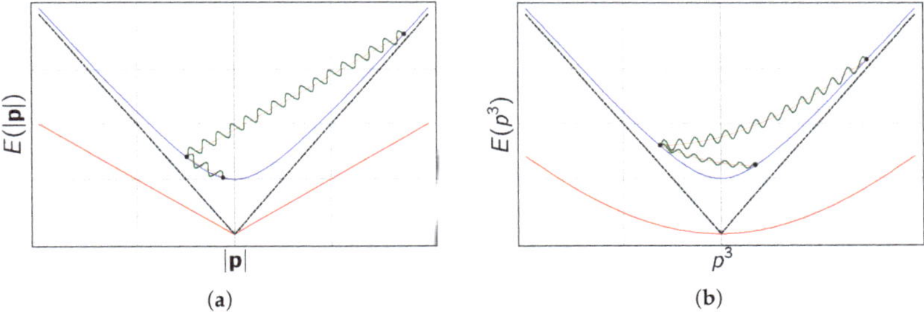

(a)   (b)

Figure 1. Standard nullcone (black, dashed), standard mass shell (blue), and modified photon dispersion relation (red) in momentum space with the momentum $\mathbf{p} = (p^1, p^2, p^3)$. Photon emissions from a standard Dirac fermion with a certain energy are indicated by green, wiggly curves. (**a**) *CPT*-even isotropic modification of the photon sector with $\tilde{\kappa}_{tr} = 3/5$. The wiggly lines run parallel to parts of the modified nullcone. In this particular example, two subsequent photon emissions are possible until the process ceases; (**b**) *CPT*-odd Maxwell–Chern–Simons (MCS) theory with $k^3_{AF} = 5m_\psi$ where only one of the two modified photon dispersion laws is illustrated. The wiggly lines run along properly shifted parts of the deformed nullcone such that the minimum is at the lower one of the two points that it connects. Two subsequent photon emissions are shown.

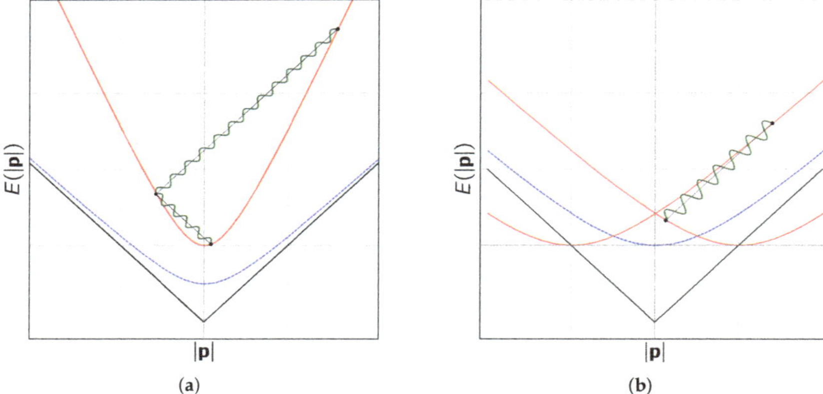

Figure 2. Standard nullcone (black), standard mass shell (blue, dashed), and modified mass shell (red) in momentum space. (**a**) isotropic spin-degenerate modification of Dirac theory with $c_{00} = -1/2$. From the initial energy of the modified fermion, two photon emissions are possible until the fermion stops radiating; (**b**) isotropic spin-nondegenerate modification of Dirac theory with $b_0 = m_\psi$. There are two distinct modified dispersion laws. Photon emission can occur from one to the other branch.

3. Properties of Vacuum Cherenkov Radiation in Minkowski Spacetime

3.1. Classical Description

The presumably first detailed study of vacuum Cherenkov radiation was carried out by Coleman and Glashow in [12]. Amongst other things, they consider an isotropic modification of the speed of light such that $c < 1$. They find that massive particles start emitting photons once their energy lies above a certain threshold energy. The rate of energy loss dE/dx is stated in the paper as well, and it goes to 0 when the particle energy approaches the threshold energy from above. Since they do not state a modified action, it is, first of all, not clear how they obtained dE/dx. A calculation of the energy loss requires knowledge of the dynamics of the process that can be inferred from a theory only. As will be explained later, the authors most probably considered the standard Feynman rules of a scalar quantum electrodynamics and a phase space modified due to Lorentz violation.

The first articles on vacuum Cherenkov radiation in the realm of the SME were written by Lehnert and Potting in 2004 [13,14]. They are based on a spacelike MCS electromagnetism coupled to standard massive particles. MCS theory is characterized by a dimensionless preferred direction ζ^μ and a mass scale m_{cs}. This framework is birefringent, i.e., it provides two distinct modified propagation modes $\ominus$ and $\oplus$ of electromagnetic waves. Independently of the observer frame chosen, $\omega_\ominus < |\mathbf{k}| < \omega_\oplus$. The properties of the dispersion laws are quite obscure in a purely timelike frame. For example, in such a frame, the mode $\oplus$ has a cusp for a vanishing momentum and a gap exists where the energy of $\ominus$ becomes complex.

To simplify the analysis, the authors study the vacuum Cherenkov process in the rest frame of the massive particle, which is possible due to observer Lorentz invariance. In the rest frame, the four-current j_μ is stationary. The solution to the four-potential can be written as an integral over momentum space involving a product of the Green's function of MCS theory and the Fourier transform $\tilde{j}_\mu$ of the four-current:

$$A^\mu(\vec{r}) = \int_{C_\omega} \frac{\mathrm{d}^3 k}{(2\pi)^3} \frac{N^{\mu\nu}(\vec{k}) \tilde{j}_\nu(\mathbf{k}) \exp(\mathrm{i}\mathbf{k} \cdot \vec{r})}{D(0, \mathbf{k})} , \tag{1}$$

where C_ω is an appropriate contour in the complex $|\mathbf{k}|$ plane and $N^{\mu\nu}$ is a tensor structure characteristic for MCS theory. The Green's function involves the inverse of the MCS dispersion equation $D(k^0, \mathbf{k})$ for $k^0 = 0$. The contour integral in the complex plane picks up contributions from the poles inside the contour only. If these poles are purely imaginary, the integrand will be exponentially damped for large distances from the trajectory. This means that the far-field vanishes whereupon there is no radiation far away from the source, which is the situation in standard electrodynamics. Hence, vacuum Cherenkov radiation occurs once there are real solutions of $D(0, \mathbf{k}) = 0$ (corresponding to spacelike dispersion relations) such that the exponential function is oscillatory. From the brief discussion of the modes of MCS theory above, we already know that such solutions exist indeed. They are given by the modes $\ominus$. The condition of real, spacelike solutions of the dispersion equation is equivalent to the existence of a refractive index $n > 1$ of the vacuum. A boost to an arbitrary observer frame provides the condition on the Cherenkov angle: $\cos\theta_c = 1/(n|\boldsymbol{\beta}|)$, where $\boldsymbol{\beta}$ is the three-velocity of the particle.

Furthermore, Lehnert and Potting found that the emitted radiation can be right-handed or left-handed polarized, dependent on the emission direction with respect to $\boldsymbol{\beta}$ and the spacelike preferred direction $\boldsymbol{\zeta}$. Using the energy-momentum tensor, it can be shown that there is no net radiated energy in the rest frame of a localized source. Furthermore, the radiated three-momentum vanishes due to the symmetry of the integrand as long as the integrand itself is regular. For real solutions of $D(0, \mathbf{k}) = 0$, the integrand is singular, in fact, and there are nonvanishing contributions to the radiated three-momentum. The latter has a backreaction on the radiating particle. Due to such recoil effects, the particle will no longer move on a geodesic, i.e., the weak equivalence principle is violated under this condition.

Further results on vacuum Cherenkov radiation at the classical level were obtained by Altschul in a series of papers. Studies on the nonbirefringent part of the CPT-even modification of Maxwell theory were carried out in [16]. These nine nonbirefringent coefficients are characterized by a symmetric, traceless matrix $\widetilde{\kappa}^{\mu\nu}$. Altschul finds that there exists a coordinate transformation between the $c^{\mu\nu}$ coefficients in the fermion sector and the $\widetilde{\kappa}^{\mu\nu}$ coefficients in the photon sector. Hence, investigating vacuum Cherenkov radiation in both sectors should produce equivalent results at first order in Lorentz violation. The possible problem of a missing frequency cutoff is identified in the computation of the total radiated energy. Therefore, integrating the radiated-energy spectrum $P(\omega)$ over the photon frequency ω may produce infinite results, as the modified theory allows for arbitrarily high ω. A hypothetical cutoff is introduced via the argument that there are possible causality problems of the theory in the vicinity of the threshold. However, we will see below that such a cutoff is obsolete when taking into account recoil effects. In the subsequent paper [17], Altschul studies additional properties of vacuum Cherenkov radiation for both MCS theory and modified Maxwell theory. He points out that MCS theory allows for a single subluminal mode only, whereas two such modes can arise in modified Maxwell theory dependent on the propagation direction. A consequence is that there is always a single Cherenkov cone only for MCS theory, whereas in modified Maxwell theory there may be two distinct cones for certain directions.

Timelike MCS theory is plagued by several problems. The energy density for the timelike sector is known to be unbounded from below. Thus, there exist runaway modes whose amplitudes increase exponentially as a function of the distance from the source. The possible origin of these issues is the gap of complex energies for the mode $\ominus$ indicated above. This gap exists for small momenta and long wavelengths, respectively, which is the motivation to study the power emitted by these long-wavelength modes in [22]. The power of the short-wavelength modes is found by a phase space estimate to be proportional to v^3 at least (with the velocity v of the radiating particle). As radiation of long wavelengths is likely to be emitted in the nonrelativistic regime and as these contributions cannot be estimated from the phasespace, they must be of $\mathcal{O}(v^2)$. The timelike MCS theory leads to a modified Ampère law where the additional term $\mathbf{J}_{\text{eff}} = 2\zeta^0\mathbf{B}$ is interpreted as a current. Altschul obtains the perturbative solution of the modified Ampère law at leading order in ζ^0 and v in a clever way. Based on this solution, the outgoing Poynting vector flux $\mathbf{S} \cdot \hat{\mathbf{e}}_r$ at $\mathcal{O}(v^2)$ is shown to be an odd function of $\cos\theta$ where θ is the polar angle. Hence, the total integral of $\mathbf{S}\cdot\hat{\mathbf{e}}_r$ over a closed surface vanishes demonstrating that the runaway modes do not contribute to vacuum Cherenkov radiation.

The latter result is generalized in [23] for the situation of a steady motion of the charge, i.e., recoil effects are neglected. The authors expand the electric and magnetic field $\mathbf{W}$ in terms of ζ^0 and the particle velocity v. They argue that each term of this expansion is either of toroidal nature, $W = W_r(r,\theta)\hat{\mathbf{e}}_r + W_\theta(r,\theta)\hat{\mathbf{e}}_\theta$, or of azimuthal form, $W = W_\phi(r,\theta)\hat{\mathbf{e}}_\phi$. The central finding is that the functions W_r, W_θ, W_ϕ for the electric field $\mathbf{E}$, the magnetic field $\mathbf{B}$, and the vector potential $\mathbf{A}$ are either even or odd with respect to $\cos\theta$. As a result of that, each combination that contributes to the outgoing Poynting flux $\mathbf{S}\cdot\hat{\mathbf{e}}_r$ provides an odd function in $\cos\theta$. Therefore, any integral of $\mathbf{S}\cdot\hat{\mathbf{e}}_r$ over a sphere vanishes and there is no net outflow of energy at all!

A possible interpretation of the latter result is that the negative energy carried away from the long-wavelength modes compensates the contributions of positive energy associated with the short-wavelength modes. Interestingly, a paper appeared very recently demonstrating this proposed cancelation explicitly [30]. The authors work in the same framework as that developed in [23]. They calculate the Fourier transform of the magnetic field at first order in ζ^0 and v. Based on this finding, all magnetic field terms of odd orders in ζ^0 can be obtained iteratively. Summing these contributions corresponds to a geometric series whose limit is an expression that is quadratic in the photon momentum and involves a singularity for the photon momentum equal to $2|\zeta^0|$. When the momentum approaches the singularity from above, there is a sign change of the whole expression demonstrating the cancelation between contributions of large momentum with those of low momentum. The singularity itself is not problematic, as its principal value can be considered formally.

3.2. Quantum Effects

The presumably first studies of the vacuum Cherenkov process in the context of quantum field theory were carried out in [15,18]. The first paper is based on a MCS theory coupled to standard Dirac fermions. Due to the issues of timelike MCS theory, an observer frame with a purely spacelike preferred direction ζ is considered. The matrix element squared of the process at tree-level is obtained based on a set of modified Feynman rules. The decay rate follows from integrating the latter result over the modified phase space. As long as the fermion under consideration has a nonvanishing momentum component $q_\parallel$ parallel to ζ, the decay rate is found to be nonzero. For $|q_\parallel| \ll m_\psi$, the decay rate depends on $q_\parallel$ in a polynomial manner, whereas for $|q_\parallel| \gg m_\psi$ it involves a logarithmic term. Such logarithmic dependencies are nonperturbative in Lorentz violation and they seem to be characteristic for the ultra-high energy regime of a birefringent theory. An additional example for a behavior of this kind will be encountered below. Furthermore, the authors find that the emitted photon is circularly polarized for large photon momenta and linearly polarized for vanishing photon momenta.

In their second paper [18], Klinkhamer and Kaufhold extend their analysis of [15]. Apart from MCS theory coupled to standard Dirac fermions, they also consider a modified scalar quantum electrodynamics, i.e., they couple MCS theory to the Lagrange density of a massive, spinless field. Including the second framework into their investigation permits them to calculate the decay rates Γ and radiated-energy rates dW/dt for spin-1/2 fermions and spinless bosons, respectively. For fermion momenta $|q_\parallel| \ll m_\psi$, spin effects are found to be highly suppressed. However, at least for large fermion momenta, they play a certain role:

$$\frac{dW_{\mathrm{scalar}}}{dt}\bigg|_{|q_\parallel| \gg m_\psi} = \frac{\alpha}{4} m_{\mathrm{cs}}|q_\parallel| + \dots\,, \qquad \frac{dW_{\mathrm{spinor}}}{dt}\bigg|_{|q_\parallel| \gg m_\psi} = \frac{\alpha}{3} m_{\mathrm{cs}}|q_\parallel| + \dots\,, \tag{2}$$

where α is the fine-structure constant.

It is interesting that these leading-order results can be obtained from a semiclassical description based on the following nontrivial refractive index of the vacuum:

$$n(\omega) = 1 + \frac{m_{\mathrm{cs}}\,\cos\theta|}{2\omega} + \dots, \tag{3}$$

with the photon energy ω and the angle θ between the photon momentum and ζ. Now, the photon spectrum for ordinary Cherenkov radiation in macroscopic media, which is provided by the Frank–Tamm formula, is reinterpreted in the context of a vacuum with the refractive index given by Equation (3). Taking the photon momentum into account, which is a quantum effect, introduces a finite maximum energy of the emitted photon. Hence, the Frank–Tamm spectrum can be integrated over the photon frequency resulting in the first radiated-energy rate of Equation (2). This finding makes sense, as Frank and Tamm did not consider the spin of the radiating particle. The latter leads to an extra contribution that adds up to the second rate of Equation (2).

A further step towards a better understanding of vacuum Cherenkov radiation in the quantum regime is achieved via [19]. In the latter reference, isotropic modified Maxwell theory based on the single coefficient $\widetilde{\kappa}_{\mathrm{tr}}$ is coupled to standard Dirac fermions. The process is found to have a threshold $E_{\mathrm{th}} \approx m_\psi / \sqrt{2\widetilde{\kappa}_{\mathrm{tr}}}$. Furthermore, the decay rate and radiated-energy rate are computed again using a set of modified Feynman rules for photons and a modified phase space. For comparison, a semiclassical analysis is performed in addition based on a modified refractive index $n(\omega) = 1 + \widetilde{\kappa}_{\mathrm{tr}} + \dots$ of the vacuum. Recoil effects lead to a finite radiated-energy rate that is found to be equal to the corresponding result in [12] for large momenta. The deduction from this finding is that Coleman and Glashow also did not take into account the particle spin. By doing so, the leading-order classical result is equal to the leading-order rate of the quantum field theory analysis including Dirac fermions.

At around the same time, the paper [20] was published. The latter also deals with isotropic modifications of both the photon and the fermion sector. In a quantum field theory analysis, the authors derive both the threshold and the decay rate for the vacuum Cherenkov process at leading order in Lorentz violation. These results are consistent with those of [19] at leading order. A constraint on isotropic Lorentz violation in the photon and electron sector follows from the assumption that an energy loss by vacuum Cherenkov radiation hides within the uncertainty of the synchrotron losses at the Large Electron-Positron Collider (LEP).

A couple of years later, Díaz and Klinkhamer refined the study carried out in [24]. Here, the parton model is used to understand effects related to a possible internal structure of the incoming fermion. In this approach, a Cherenkov photon can be emitted from a single parton inside a fermion, i.e., neutrons can radiate vacuum Cherenkov radiation, as well. The authors determine the emission rate by an electrically charged parton carrying the fraction xp of the incoming fermion momentum p. The maximum photon energy is found to be proportional to the fraction x. Hence, Cherenkov emission is suppressed for small momentum fractions, which practically excludes this process for the sea quarks. Another interesting property is that emissions of very hard photons may lead to momentum transfers large enough to destroy the initial hadron and to produce new ones in the final state. Finally, the total power radiated by a composite proton and neutron, respectively, is determined based on very recent data on parton distribution functions. The rates for both an initial proton and neutron are suppressed by around one order of magnitude in comparison to the rate in [19] for a structureless Dirac fermion. This result does not change the constraint obtained in [19] though.

A quantum field theory analysis that complements the classical studies of [22,23,30] for vacuum Cherenkov radiation in timelike MCS theory is carried out in [27]. Complex energies of the mode $\ominus$ can be avoided by introducing a nonvanishing photon mass. Note that the current experimental limit on a photon mass is $m_\gamma < 1 \times 10^{-27}$ GeV [31], which is many orders of magnitudes larger than the best constraint on the controlling coefficient under consideration, which is $\zeta^0 \lesssim 10^{-43}$ GeV [32]. A nonzero photon mass also allows for a consistent quantization procedure whereupon the decay rate for vacuum Cherenkov radiation can be calculated in quantum field theory. The process has a threshold and the decay rate for initial fermion momenta much larger than this threshold is characterized by two different regimes. There is a range of momenta where it is suppressed quadratically by Lorentz violation and increases quadratically with the fermion momentum. For very large momenta, it is only linearly suppressed by Lorentz violation and grows linearly with momentum.

After the big amount of investigations of vacuum Cherenkov radiation with modified photons, time was ready to study the vacuum Cherenkov process for Lorentz-violating fermions in [28]. The coordinate transformation moving Lorentz violation between the photon and fermion sector [16] only covers the CPT-even, nonbirefringent c coefficients for fermions. Such a transformation is neither known for the CPT-odd a, b, e, f, and g coefficients nor for the CPT-even, birefringent d, H coefficients, which is why such an analysis is definitely reasonable. Note also that for the birefringent b, d, H, and g coefficients, there are two fermion modes that will be denoted as $\oplus_X$ and $\ominus_X$ where the index X stands for a controlling coefficient. This property leads to interesting new effects that do not have a counterpart in the photon sector. The base of the analysis performed in [28] are the modified Feynman rules for external fermion legs developed in the technical work [33]. Due to calculational complexities, decay rates and radiated-energy rates are obtained numerically for a broad range of coefficients. The essential properties are that the processes for the c, d, e, f, and g coefficients have thresholds. Furthermore, for large incoming fermion momenta, the decay rates for the c, d coefficients are linearly suppressed in Lorentz violation, whereas the decay rates for e, f, and g are quadratically suppressed. The processes $\oplus_{d,g} \rightarrow \oplus_{d,g} + \gamma$ are allowed for a certain sign of the controlling coefficients and $\ominus_{d,g} \rightarrow \ominus_{d,g} + \gamma$ occur for the opposite sign. Analog decays for the a, b, and H coefficients are found to be forbidden. However, the birefringent nature of the spin-nondegenerate coefficients opens up the possibility of spin–flip processes $\oplus_{b,d,H,g} \rightarrow \ominus_{b,d,H,g} + \gamma$ and the analog ones with $\oplus$ and $\ominus$ interchanged. These decays have peculiar properties. First, they do not have a threshold and, second, their decay rates are highly suppressed by several powers of the controlling coefficients in contrast to the spin-conserving decays. Furthermore, the decay rates involve logarithmic terms for large momenta similarly to the results for MCS theory obtained in [15].

3.3. Radiation of Particles Other Than Photons

A plethora of alternative Cherenkov-type processes is conceivable in Lorentz-violating theories. The radiating particle does not necessarily have to be a massive fermion and the radiated particles need not inevitably be photons. One of the first articles in this realm was published by Altschul [26]. Here, an effective theory for Lorentz-violating pions is considered coupled to standard photons. In this framework, high-energy photons can lose energy by radiating pions where the process involves a threshold. At leading order in Lorentz violation, the threshold has parallels with the analog expression for nonbirefringent modified Maxwell theory in [16]. After all, both frameworks are quite similar from a merely kinematic point of view. As the direct calculation of the decay rate is cumbersome, an estimate of the rate is obtained from the prominent process $\Gamma(\pi^0 \rightarrow 2\gamma)$.

A further interesting Cherenkov-type process is explored in [29]. The authors consider a Lorentz-violating modification of the free W-boson sector. Such a term enables a Cherenkov-type process in vacuo where a fermion of mass $m_1 < m_W$ decays into a fermion of mass $m_2 < m_W$ under the emission of a W boson. An analysis in the rest frame of the incoming particle is feasible. It reveals that the process is energetically possible for a quite large controlling coefficient in the rest frame showing that this frame must be strongly boosted with respect to a concordant frame where Lorentz violation is supposed to be small. Analogous to [24], the authors perform a sophisticated parton model study based on the elementary process mentioned before. Hence, the decay rate of W-boson emission of a single quark is convoluted with appropriate parton distribution functions. As the probability of finding quarks with a large momentum fraction xp steeply decreases for $x \mapsto 1$, the decay rate of a proton right at the threshold region is considerably suppressed in comparison to the decay rate of the elementary process.

After the detection of superluminal neutrinos had been announced by the OPERA collaboration in 2011, Cohen and Glashow quickly published a paper studying Cherenkov-type radiation of electron-positron pairs by neutrinos via virtual Z bosons [21]. Based on the value for Lorentz violation in the neutrino sector announced by OPERA, they showed that neutrinos would quickly lose their energy via this Cherenkov process making it impossible for them to arrive at the Gran Sasso laboratory with the energy that was measured by OPERA. Few months later, OPERA announced that errors had occurred in the course of the measurement process rendering their results incorrect.

4. Cherenkov-Type Radiation in Modified Gravity

In this review, special emphasis shall be put on Cherenkov-type processes in theories of modified gravity. Before doing so, it is reasonable to recall certain properties of General Relativity that are crucial in the context of Lorentz violation. We will refer to the Einstein equations without the cosmological constant that are given by

$$G_{\mu\nu} = 8\pi G_N (T_M)_{\mu\nu} \, , \tag{4a}$$

$$G_{\mu\nu} = R_{\mu\nu} - \frac{R}{2} g_{\mu\nu} \, . \tag{4b}$$

Here, $G_{\mu\nu}$ is the Einstein tensor that decomposes into the Ricci tensor $R_{\mu\nu}$ and the Ricci scalar R. The spacetime metric is given by $g_{\mu\nu}$ and $(T_M)_{\mu\nu}$ stands for the energy-momentum tensor of matter. The coupling constant of gravity and matter corresponds to Newton's constant G_N.

4.1. Lorentz Violation in Gravity

General Relativity exhibits invariance under diffeomorphisms and under general coordinate transformations where these concepts are equivalent. A diffeomorphism is a differentiable map of a manifold onto itself whose inverse has the same properties. Diffeomorphism invariance can also be understood as a gauge symmetry of the theory. It is possible to choose a gauge fixing condition appropriately to remove four of the ten independent components of the spacetime metric. Hence, only six metric components may possibly be physical. Since one set of Bianchi identities of Riemannian geometry provides $D_\mu G^{\mu\nu} = 0$ with the covariant derivative D_μ, an alternative argument is that the latter four identities eliminate four spacetime metric components. Thus, we see that the Bianchi identities are closely connected to diffeomorphism invariance. Diffeomorphisms in Minkowski spacetime simply correspond to translations. After all, translations map Minkowski spacetime onto itself.

The global Lorentz invariance of Special Relativity is lost in General Relativity. However, there still exists the concept of local Lorentz invariance. To understand what that means, it is useful to formulate General Relativity with the vierbein approach. The vierbein (tetrad) allows for transforming between the coordinate basis of the spacetime manifold and the basis of a local inertial (freely falling) reference frame at each spacetime point. The great advantage of dealing with the vierbein formalism is that it allows us to introduce Lorentz invariance as a local concept, i.e., a symmetry that is valid in a local frame [34].

In the context of gravity, matter in curved spacetimes is described by minimally coupling appropriate fields to the Einstein–Hilbert action. This means that the Minkowski metric has to be replaced by a curved spacetime metric wherever it occurs. Furthermore, ordinary partial derivatives are replaced by covariant derivatives on the manifold. Dependent on whether the metric or vierbein formalism is used, the determinant of the metric and the vierbein, respectively, has to be taken into account in the volume element of the action. Furthermore, the vierbein formalism is indispensable when introducing Dirac fermions into the theory due to their additional spinor structure. As it is clear how to deal with the spinor structure in a local inertial frame, the vierbein formalism extends the concept of a spinor from such a local frame to the frame of spacetime coordinates.

There are two types of local Lorentz transformations: observer and particle transformations. With matter coupled to General Relativity, it is easier to clarify their implications. Observer Lorentz transformations correspond to rotations and boosts of a local reference frame, whereas particle Lorentz transformations deal with rotations and boosts of matter fields in such local frames. (Note that the SME community avoids the terms passive and active transformations that are actually used in the literature. One of the reasons is that an active transformation may be supposed to transform both the background field and the matter fields leaving the coordinates unchanged. In contrast, a particle transformation touches the matter field only, whereas the background field remains fixed.) The term general coordinate transformations already suggests that this type of transformations only changes the choice of coordinates of the base manifold. On the contrary, diffeomorphisms transform the matter fields including the metric tensor, which in the context of General Relativity is considered as a field as well. Therefore, general coordinate transformations as well as observer Lorentz transformations merely change the description of the physics, but leave the physics itself untouched. On the other hand, both diffeomorphisms and particle Lorentz transformations act on the physical fields though. Diffeomorphisms and general coordinate transformations are equivalent in General Relativity and so are particle and observer Lorentz transformations. However, when including Lorentz violation into the theory, these concepts have to be distinguished from each other carefully, cf. Figures 3 and 4 for an illustration on a curved manifold and its tangent plane, respectively.

The gravitational part of the SME is developed in [34] and its formulation is based on the vierbein formalism. In the gravitational SME, the conceptual differences between diffeomorphisms and general coordinate transformations as well as particle and observer Lorentz transformations become evident. The SME introduces background fields that break local particle Lorentz invariance, i.e., there may be one or several preferred directions in each freely falling reference frame. The background field does not transform like a tensor anymore under particle Lorentz transformations, but it remains fixed. However, there is still the freedom to choose coordinates within these frames as desired. A violation of local particle Lorentz invariance leads to a violation of diffeomorphism invariance and vice versa. The reason is that any coordinate-dependent vector field on a curved manifold has a homogeneous counterpart in a local inertial reference frame at each point of the manifold and vice versa. As local observer Lorentz invariance is still granted, so is invariance under general coordinate transformations.

In the gravitational SME, the minimal matter-sector SME contributions, which were originally formulated in Minkowski spacetime, are minimally coupled to a curved spacetime. The vierbein must be accompanied by an object known as the spin connection granting that the covariant derivative of the vierbein vanishes. The spin connection and the affine connection (Christoffel symbols) are not independent from each other. Apart from Lorentz violation in the matter sector, the gravitational SME describes Lorentz violation in the pure-gravity sector. The dimension-4 contributions incorporate a single Lorentz-invariant term multiplied by the curvature scalar, a symmetric matrix of nine nonbirefringent coefficients contracted with the Ricci tensor, and 10 birefringent coefficients contracted with the Weyl tensor, which is the totally traceless Riemann tensor [34].

In addition to the properties outlined above, there is a further highly subtle statement on the origin of Lorentz violation in gravity. First of all, the Lorentz-violating background fields in Minkowski spacetime are usually taken as independent from the spacetime coordinates, which implies energy-momentum conservation. This choice is reasonable from a physical perspective, since a constant background can be considered as the dominant contribution and a part dependent on the spacetime coordinates is interpreted as a correction. In the context of a curved spacetime, however, it is not clear what "constant" means. A first justified thought may be to require that the covariant derivative of a background field vanish. However, there are only few manifolds that permit such a requirement and those are not that interesting in the context of gravity. Anyhow, background fields in gravity will have some dependence on the spacetime coordinates. Therefore, Lorentz violation in the matter sector leads to modifications of the conservation law of the energy-momentum tensor where the latter now includes nonvanishing spacetime derivatives of the controlling coefficients. However, the identity $D_\mu G^{\mu\nu} = 0$ is still valid even in the presence of Lorentz violation, as it simply follows from the underlying Riemannian framework. Hence, there is a clash when including Lorentz violation into a gravity theory by hand (explicit Lorentz violation) [34]. The way out of this no-go theorem is to consider spontaneous Lorentz violation, i.e., Lorentz-violating background fields that have a dynamical origin and satisfy field equations on their own. Therefore, the action of the gravitational SME must also include terms that describe possible mechanisms providing a way of how to break local Lorentz invariance spontaneously.

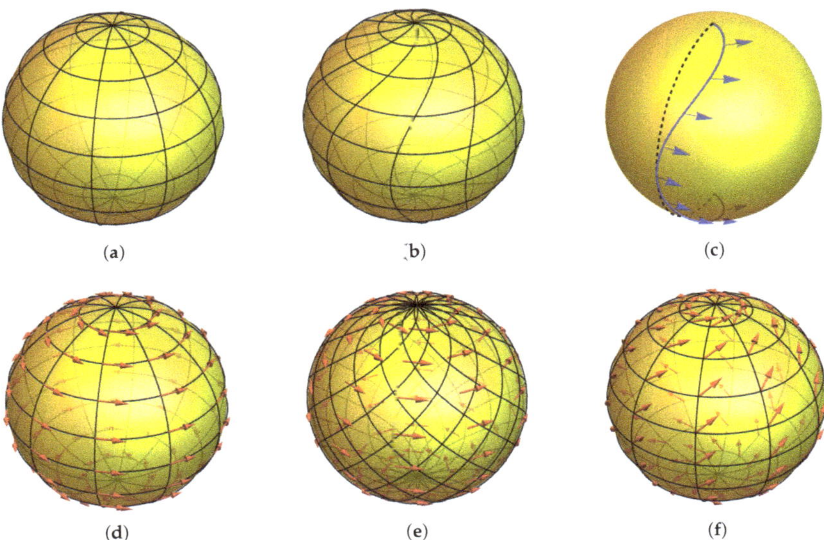

Figure 3. (**a**) two-sphere (S^2) parameterized in terms of the polar angle ϑ and the azimuthal angle φ. The coordinate lines of constant ϑ are circles parallel to the equator where the coordinate lines of constant φ are great circles linking the north and south poles with each other; (**b**) illustration of a general coordinate transformation for S^2 that deforms the original coordinate lines; (**c**) illustration of a diffeomorphism for S^2. To distinguish it from a general coordinate transformation, the unchanged coordinate lines are not shown. The graphics demonstrates how an original geodesic connecting the north and the south pole gets deformed by the diffeomorphism; (**d**) S^2 endowed with an additional vector field that could correspond to a matter field in physics; (**e**) general coordinate transformation applied to S^2. The coordinate lines are deformed, but the vector field itself stays untouched. As the coordinates change, however, the explicit representation of the vector field changes as well; (**f**) diffeomorphism applied to S^2 with an intrinsic vector field. In contrast to before, the coordinates remain unchanged, but the vector field transforms in a nontrivial manner.

An explicit violation of Lorentz invariance in a theory of modified gravity is analogous to an explicit violation of gauge invariance in a gauge theory. The latter is obviously considered as destructive for the fate of a gauge theory because it leads to a breakdown of unitarity, the production of unphysical modes, etc. In contrast to that, a spontaneous violation of gauge invariance is a tremendously useful physical mechanism capable of explaining a multitude of effects both in particle physics and in condensed-matter physics. For a spontaneously violated gauge symmetry, the underlying theory is still perfectly gauge-invariant. It is just the ground state of the theory that randomly picks a certain gauge configuration, cf. the vacuum expectation value of the Higgs field in the Higgs mechanism. In the context of Lorentz violation, a background field emerges as vacuum expectation values of a tensor-valued fundamental field. Hence, the underlying fundamental theory is Lorentz invariant, but the ground state picks certain preferred directions in spacetime. As the background field is dynamic, picking a definite ground state is accompanied by fluctuations, i.e., both Nambu–Goldstone modes and massive modes are created. The first correspond to small fluctuations of the preferred directions around the direction chosen by the vacuum expectation values. The second are associated with fluctuations of the magnitude of controlling coefficients.

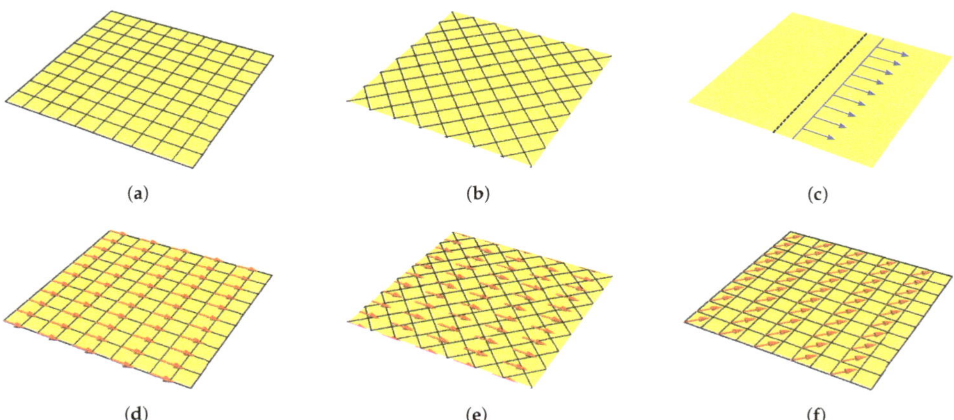

Figure 4. (a) tangent plane of S^2 with coordinate lines shown; (b) observer Lorentz transformation rotating the coordinate lines; (c) diffeomorphism (translation) that simply maps the points of a geodesic (straight line) to the points of a parallel geodesic. The unmodified coordinate lines are omitted for clarity; (d) tangent plane endowed with a vector field; (e) observer Lorentz transformation that changes the coordinate lines, but leaves the vector field untouched; (f) particle Lorentz transformation that induces a nontrivial transformation of the vector field. The coordinate lines are not modified.

4.2. Gravitational Vacuum Cherenkov Radiation

After reviewing some of the crucial implications of Lorentz violation in gravity, we are now ready to discuss the characteristics of the Cherenkov-type process in gravity that is investigated in [25]. In their paper, Kostelecký and Tasson consider the nonbirefringent coefficients of the pure-gravity sector. In contrast to [34], all operators of even mass dimension $d \geq 4$ are taken into account. Therefore, this analysis is currently the only one including contributions from operators of higher mass dimension. For the purpose of studying Cherenkov-type processes in gravity, it suffices to restrict their analysis to linearized gravity, i.e., the metric perturbation $h_{\mu\nu}$ is neglected at second and higher orders in the field equations [35,36]. The metric perturbation has 10 components out of which four are gauge degrees of freedom, as discussed before. Another four components are auxiliary and two are physical ones that propagate. The latter two correspond to the polarization modes of a gravitational wave.

Lorentz violation is restricted to first order as well. The corresponding coefficients are assumed to be constant and small. In principle, they are put in by hand, which is why Lorentz invariance breaking is explicit in this treatment. However, the coefficients considered are chosen such that gauge symmetry (diffeomorphism invariance) is restored at linear order [35]. (In [36], the claim for gauge invariance is relaxed and sets of coefficients are included that violate gauge invariance.) Therefore, the number of polarization modes of a gravitational wave remains fixed and merely Lorentz-violating perturbations of the standard polarizations are expected. Both the gauge and auxiliary components are removed by appropriately adapted gauge fixing conditions. These are modifications of the standard

Hilbert gauge and transverse traceless gauge conditions that are usually employed in the context of linearized gravity.

The resulting Einstein equations have the form of a wave equation for the metric perturbation $h_{\mu\nu}$. The solutions of these equations are interpreted as gravitational waves perturbed by Lorentz violation. A description of this kind introduces a nontrivial refractive index of the vacuum. In the current context, the latter is a refractive index for gravitational waves instead of electromagnetic waves. At linear order in Lorentz violation, the refractive index involves a sum over all possible Lorentz-violating contributions of even mass dimension $d \geq 4$ where the coefficients for $d > 4$ are contracted with additional wave four-vectors. Therefore, the modified gravity theory exhibits anisotropy, but no dispersion for $d = 4$, whereas, for $d > 4$, both anisotropy and dispersion play a role.

Within a semiclassical approach, a gravitational wave is quantized in the sense that it carries a momentum in analogy to a photon resulting from the quantization of electromagnetic waves. Quanta of the linearized gravitational field are introduced as well, and they are known as gravitons. Within this semiclassical treatment of linearized gravity, Kostelecký and Tasson derive modified perturbative Feynman rules for distinct particle sectors. They consider graviton emission from spinless bosons, photons, and Dirac fermions. Each of these vertices is proportional to the square root of Newton's constant G_N, i.e., the matrix element squared is directly proportional to G_N and quadratically suppressed by Lorentz violation, in addition.

The notion of conserved quantities along geodesics is a very subtle one in General Relativity. It requires some knowledge of isometries of the underlying curved spacetime metric encoded in the Killing vectors. However, local energy-momentum conservation directly follows from Einstein's equations combined with $\partial_\mu G^{\mu\nu} = 0$. Since interactions occur locally in the language of perturbation theory and Feynman rules, energy-momentum conservation at each vertex is granted. Hence, kinematic arguments such as those outlined in Section 2 also apply.

The radiated-energy rate follows from integrating the matrix element squared over an appropriately modified phase space. Due to the nontrivial refractive index of the vacuum, particles may travel faster than the phase velocity of gravity in vacuo, rendering graviton emission possible. This process is of Cherenkov-type and can occur for all known particles, as they all interact with gravity. Analogous to conventional vacuum Cherenkov radiation, a Cherenkov angle can be introduced as the angle between the propagation direction of the radiating particle and the gravitons radiated. The Cherenkov angle is defined only above a certain velocity of the incoming particle demonstrating that gravitational Cherenkov radiation is a threshold process. Another interesting property is that energy-momentum conservation introduces an upper cutoff for the energy of the emitted gravitons. The latter corresponds approximately to the incoming particle momentum. Thus, the phase space integral is rendered finite without any issues. Recall the analog behavior found for vacuum Cherenkov radiation in Minkowski spacetime [15,18,19].

In the ultrarelativistic limit, integration over phase space for any of the three processes considered simply delivers a factor $F^w(d)$ that depends on the mass dimension of the Lorentz-violating operators. Here, $w = \{\phi, \psi, \gamma\}$ for scalars, Dirac fermions, and photons, respectively. In general, the energy loss per time has the form

$$\frac{\mathrm{d}E}{\mathrm{d}t} = -F^w(\hat{a})G_N(s^{(d)})^2 |\mathbf{p}|^{2d-4} , \tag{5}$$

with the incoming particle momentum $\mathbf{p}$ and the Lorentz-violating operator $s^{(d)}$, which is a direction-dependent combination of coefficients of mass dimension d. The factors $F^w(d)$ have the property that $F^\phi < F^\psi < F^\gamma$, where $F^\phi \sim 16/d$, $F^\psi \sim 8$, and $F^\gamma \sim 16d$ for $d \mapsto \infty$. Hence, a photon loses more energy by gravitational Cherenkov radiation than a Dirac fermion and a Dirac fermion more than a spinless particle. The reason for this behavior is to be found in the spin of the radiating particle. A graviton is a spin-2 excitation, i.e., its spin projection $S_{\hat{\xi}}$ with respect to some quantization axis $\hat{\xi}$ can take each value out of the set $S_{\hat{\xi}} = \{-2, -1, 0, 1, 2\}$. Hence, the emission of gravitons with $S_{\hat{\xi}} = \pm 2$ is only possible for photons of polarizations $\lambda_+ = 1$ and $\lambda_- = -1$, respectively. A complete flip of the polarization $\lambda_{\pm} \mapsto \lambda_{\mp}$ during the process can deliver $S_{\zeta} = \pm 2$ necessary to produce such gravitons. For Dirac fermions, only gravitons with $S_{\zeta} = \{-1, 0, 1\}$ can be emitted where for scalars $S_{\zeta} = 0$ is the remaining possibility. Since there are less decay channels for Dirac fermions in comparison to photons, energy loss of fermions is suppressed. A similar argument holds for scalars in comparison to fermions.

The main objective of [25] is to place stringent bounds on Lorentz violation in the pure-gravity sector. From the behavior of $F^{w}(d)$, we see that high-energy cosmic photons have the largest potential of doing so. However, as cosmic fermions with much higher energy are observed, the analysis in [25] is based on fermions. To simplify it, the authors consider operators of a particular mass dimension at a time. To constrain isotropic Lorentz violation, a single cosmic-ray event is sufficient. However, to carry out the same for a whole set of anisotropic coefficients, several such events coming from different directions are needed. It is convenient to state these directions in terms of azimuthal and polar angles on the celestial sphere. Hence, it also makes sense to expand the controlling coefficients in spherical harmonics. A considerable compilation of high-energy cosmic-ray events is chosen. To obtain conservative constraints, Kostelecký and Tasson base their analysis on iron nuclei where a graviton is assumed to be emitted by one of the fermionic partons. In contrast to [24], they do not consider parton distribution functions explicitly. Instead, the fermionic parton emitting gravitons is taken to have 10% of the total incoming cosmic-ray energy. A set of bounds for a given d is obtained by a maximizing procedure to fulfill the requirement that Lorentz violation can only be as large as to not obstruct cosmic rays with a certain energy from arriving on Earth.

By doing so, it is possible to constrain Lorentz violation for $d = 4$, 6, and 8. The isotropic constraints are one-sided, whereas the anisotropic ones are two-sided. The rough order of magnitude of the constraints is 10^{-13} for $d = 4$, $10^{-29}\,\mathrm{GeV}^{-2}$ for $d = 6$, and $10^{-45}\,\mathrm{GeV}^{-4}$ for $d = 8$. These bounds are impressive and are currently the best ones in the pure-gravity sector. As constraints on Lorentz violation in the neutrino sector are the tightest ones, they can serve as a reference for comparisons. Very recently, the ICECUBE collaboration published a paper including novel neutrino bounds obtained from the observation of ultra-high energy neutrinos [37]. The constraints for the dimension-6 neutrino coefficients are better than those in the gravity sector by several orders of magnitude, whereas the sensitivities for the dimension-8 coefficients can compete with each other. If ultrahigh-energy protons can clearly be identified as primaries of cosmic-ray air showers, the gravity sector bounds have a tremendous potential of improvement.

5. Conclusions

The intention of the current review was to provide a summary of the crucial concepts of vacuum Cherenkov radiation developed over the past 20 years. At this point, it makes sense to take a look at the set of constraints on SME coefficients obtained from the absence of vacuum Cherenkov radiation, which are compiled in Table 1. Most of the bounds are on dimensionless coefficients except for the constraints in the W boson and the graviton sector partially. A subset of the bounds in the graviton sector are the only ones on nonminimal coefficients showing that most studies on vacuum Cherenkov radiation so far have been restricted to the minimal SME. The birefringent coefficients of both MCS theory and modified Maxwell theory have never been constrained via the absence of vacuum Cherenkov radiation. The reason is that this type of coefficients can be much more tightly bounded by the absence of vacuum birefringence.

The isotropic c coefficients and the isotropic $\widetilde{\kappa}_{\mathrm{tr}}$ are linked by the coordinate transformation between the photon and the matter sector found in [16]. Furthermore, the f coefficients squared correspond to c coefficients [38]. Hence, bounds are usually placed on combinations of these coefficients. The related ones in the proton sector are quite stringent, whereas the bound in the electron sector is weaker. As free electrons are not expected as a component of high-energy cosmic rays, the latter constraint is based on the maximum energy reached at LEP [20]. The constraints for the e and g coefficients are also weaker, since the related threshold energies depend inversely on these coefficients instead of via an inverse square root of the coefficient [28].

As the bound on Lorentz violation in the minimal gravity sector is dimensionless, it can be compared to the remaining dimensionless constraints. Although the results obtained in the pure-gravity sector are currently the best ones, they are weaker than the best dimensionless constraints in the photon and fermion sector. The reason is that gravity involves the dimensionful coupling constant G_N in comparison to the other three fundamental interactions. Hence, the energy loss of Equation (5) is directly proportional to G_N and needs two additional powers of the incoming fermion momentum to cancel the inverse mass dimensions of G_N. The additional powers of the momentum are not enough to compensate for the smallness of G_N. Furthermore, Equation (5) is not just linear in Lorentz violation, but depends quadratically on it.

The future perspective in this interesting subfield is to continue studying Cherenkov-type processes in vacuo based on Lorentz violation of the nonminimal SME. With the detection of cosmic-ray events of higher energy and an improved identification of primaries in air showers, there is a great potential of even better tests of the very foundations of both the Standard Model and General Relativity.

Table 1. Compilation of constraints on SME coefficients obtained from the absence of Cherenkov-type radiation in the vacuum. Most of the constraints can be found in the data tables [32]. As the number of constraints in the gravity sector is extensive, approximate constraints on groups of coefficients are given where $0 < j \leq d - 2$ (with the mass dimension d) and $0 \leq m \leq j$.

Sector	Constraint	Reference
Proton	$-5 \times 10^{-23} < \mathring{c}^{\text{UR}(4)}$	[12]
Proton, Photon	$\widetilde{\kappa}_{\text{tr}} - (4/3)c_{00}^{(4)\text{p}} < 6 \times 10^{-20}$	[19]
Electron, Photon	$\widetilde{\kappa}_{\text{tr}} - (4/3)c_{00}^{(4)\text{e}} < 1.2 \times 10^{-11}$	[20]
Gravity	$\bar{s}_{00}^{(4)} > -3 \times 10^{-14}$	[25]
Gravity	$|\bar{s}_{jm}^{(4)}| \lesssim 10^{-13}$	[25]
Gravity	$\bar{s}_{00}^{(6)} < 2 \times 10^{-31}\,\text{GeV}^{-2}$	[25]
Gravity	$|\bar{s}_{jm}^{(6)}| \lesssim 10^{-29}\,\text{GeV}^{-2}$	[25]
Gravity	$\bar{s}_{00}^{(8)} > -7 \times 10^{-49}\,\text{GeV}^{-4}$	[25]
Gravity	$|\bar{s}_{jm}^{(8)}| \lesssim 10^{-45}\,\text{GeV}^{-4}$	[25]
Pion	$\delta^{\pi} > -7 \times 10^{-13}$	[26]
Quark, Photon	$-3 \times 10^{-23} \leq c_{00}^{(4)\text{u}} - (3/4)\widetilde{\kappa}_{\text{tr}} - (3/8)(f_0^{(4)\text{u}})^2$	[28]
Quark	$|\mathring{a}_{00}^{(4)\text{u}}| < 3 \times 10^{-23}$	[28]
Quark	$|\mathring{c}_0^{(4)\text{u}}| < 9 \times 10^{-12}$	[28]
Quark	$|\mathring{g}_1^{(4)\text{u}}| < 9 \times 10^{-12}$	[28]
W boson	$|(k_1)^{\mu}| < 1.7 \times 10^{-8}\,\text{GeV}$	[29]
W boson	$|(k_2)^{\mu}| < 1.1 \times 10^{-7}\,\text{GeV}$	[29]

Funding: This research was funded by CNPq grant number 421566/2016-7 and by FAPEMA grant number 01149/17.

Acknowledgments: It is a pleasure to thank the Brazilian agencies CNPq and FAPEMA for their financial support. Furthermore, the author is indebted to the two anonymous referees for useful comments on the submitted version of the paper.

Conflicts of Interest: The author declares no conflict of interest.

References

1. Cherenkov, P.A. Visible emission of clean liquids by action of γ radiation. *Dokl. Akad. Nauk SSSR* **1934**, *2*, 451.
2. Tamm, I.E.; Frank, I.M. Coherent radiation of fast electrons in a medium. *Dokl. Akad. Nauk SSSR* **1937**, *14*, 107.
3. Jelley, J.V. *Čerenkov Radiation and Its Applications*; Pergamon Press: London, UK, 1958.

4. Kostelecký, V.A.; Samuel, S. Spontaneous breaking of Lorentz symmetry in string theory. *Phys. Rev. D* **1989**, *39*, 683. [CrossRef]

5. Gambini, R.; Pullin, J. Nonstandard optics from quantum space-time. *Phys. Rev. D* **1999**, *59*, 124021. [CrossRef]

6. Colladay, D.; Kostelecký, V.A. Lorentz-violating extension of the standard model. *Phys. Rev. D* **1998**, *58*, 116002. [CrossRef]

7. Colladay, D.; Kostelecký, V.A. *CPT* violation and the standard model. *Phys. Rev. D* **1997**, *55*, 6760. [CrossRef]

8. Sommerfeld, A. On the theory of electrons I. *Gött. Nachr.* **1904**, *2*, 99.

9. Sommerfeld, A. On the theory of electrons II. *Gött. Nachr.* **1904**, *2*, 363.

10. Sommerfeld, A. On the theory of electrons III. *Gött. Nachr.* **1905**, *3*, 201.

11. Beall, E.F. Measuring the gravitational interaction of elementary particles. *Phys. Rev. D* **1970**, *1*, 961. [CrossRef]

12. Coleman, S.R.; Glashow, S.L. Cosmic ray and neutrino tests of special relativity. *Phys. Lett. B* **1997**, *405*, 249. [CrossRef]

13. Lehnert, R.; Potting, R. The Cerenkov effect in Lorentz-violating vacua. *Phys. Rev. D* **2004**, *70*, 125010. [CrossRef]

14. Lehnert, R.; Potting, R. Vacuum Cerenkov radiation. *Phys. Rev. Lett.* **2004**, *93*, 110402. [CrossRef] [PubMed]

15. Kaufhold, C.; Klinkhamer, F.R. Vacuum Cherenkov radiation and photon triple-splitting in a Lorentz-noninvariant extension of quantum electrodynamics. *Nucl. Phys. B* **2006**, *734*, 1–23. [CrossRef]

16. Altschul, B. Vacuum Cerenkov radiation in Lorentz-violating theories without *CPT* violation. *Phys. Rev. Lett.* **2007**, *98*, 041603. [CrossRef] [PubMed]

17. Altschul, B. Cerenkov Radiation in a Lorentz-violating and birefringent vacuum. *Phys. Rev. D* **2007**, *75*, 105003. [CrossRef]

18. Kaufhold, C.; Klinkhamer, F.R. Vacuum Cherenkov radiation in spacelike Maxwell–Chern–Simons theory. *Phys. Rev. D* **2007**, *76*, 025024. [CrossRef]

19. Klinkhamer, F.R.; Schreck, M. New two-sided bound on the isotropic Lorentz-violating parameter of modified-Maxwell theory. *Phys. Rev. D* **2008**, *78*, 085026. [CrossRef]

20. Hohensee, M.A.; Lehnert, R.; Phillips, D.F.; Walsworth, R.L. Limits on isotropic Lorentz violation in QED from collider physics. *Phys. Rev. D* **2009**, *80*, 036010. [CrossRef]

21. Cohen, A.G.; Glashow, S.L. Pair creation constrains superluminal neutrino propagation. *Phys. Rev. Lett.* **2011**, *107*, 181803. [CrossRef] [PubMed]

22. Altschul, B. Absence of long-wavelength Cerenkov radiation with isotropic Lorentz and *CPT* violation. *Phys. Rev. D* **2014**, *90*, 021701. [CrossRef]

23. Schober, K.; Altschul, B. No vacuum Cerenkov radiation losses in the timelike Lorentz-violating Chern-Simons theory. *Phys. Rev. D* **2015**, *92*, 125016. [CrossRef]

24. Díaz, J.S.; Klinkhamer, F.R. Parton-model calculation of a nonstandard decay process in isotropic modified Maxwell theory. *Phys. Rev. D* **2015**, *92*, 025007. [CrossRef]

25. Kostelecký, V.A.; Tasson, J.D. Constraints on Lorentz violation from gravitational Čerenkov radiation. *Phys. Lett. B* **2015**, *749*, 551–559. [CrossRef]

26. Altschul, B. Cerenkov-like emission of pions by photons in a Lorentz-violating theory. *Phys. Rev. D* **2016**, *93*, 105007. [CrossRef]

27. Colladay, D.; McDonald, P.; Potting, R. Cherenkov radiation with massive, *CPT*-violating photons. *Phys. Rev. D* **2016**, *93*, 125007. [CrossRef]

28. Schreck, M. Vacuum Cherenkov radiation for Lorentz-violating fermions. *Phys. Rev. D* **2017**, *96*, 095026. [CrossRef]

29. Colladay, D.; Noordmans, J.P.; Potting, R. Cosmic-ray fermion decay by emission of on-shell W bosons with *CPT* violation. *Phys. Rev. D* **2017**, *96*, 035034. [CrossRef]

30. DeCosta, R.; Altschul, B. Mode analysis for energetics of a moving charge in Lorentz- and *CPT*-violating electrodynamics. *Phys. Rev. D* **2018**, *97*, 055029. [CrossRef]

31. Particle Data Group. Review of Particle Physics. *Phys. Rev. D* **2018**, *98*, 030001. [CrossRef]

32. Kostelecký, V.A.; Russell, N. Data tables for Lorentz and *CPT* violation. *Rev. Mod. Phys.* **2011**, *83*, 11. [CrossRef]

33. Reis, J.A.A.S.; Schreck, M. Lorentz-violating modification of Dirac theory based on spin-nondegenerate operators. *Phys. Rev. D* **2017**, *95*, 075016. [CrossRef]
34. Kostelecký, V.A. Gravity, Lorentz violation, and the standard model. *Phys. Rev. D* **2004**, *69*, 105009. [CrossRef]
35. Kostelecký, V.A.; Mewes, M. Testing local Lorentz invariance with gravitational waves. *Phys. Lett. B* **2016**, *757*, 510–514. [CrossRef]
36. Kostelecký, V.A.; Mewes, M. Lorentz and diffeomorphism violations in linearized gravity. *Phys. Lett. B* **2018**, *779*, 136–142. [CrossRef]
37. IceCube Collaboration. Neutrino interferometry for high-precision tests of Lorentz symmetry with IceCube. *Nat. Phys.* **2018**, *14*, 961. [CrossRef]
38. Altschul, B. Eliminating the CPT-odd f coefficient from the Lorentz-violating standard model extension. *J. Phys. A* **2006**, *39*, 13757. [CrossRef]

Article

The (A)symmetry between the Exterior and Interior of a Schwarzschild Black Hole

Pawel Gusin [1], Andy Augousti [2], Filip Formalik [1] and Andrzej Radosz [1,*]

[1] Department of Quantum Technolgies, Wroclaw University of Science and Technology,
    50-370 Wroclaw, Poland; Pawel.Gusin@pwr.edu.pl (P.G.); Filip.Formalik@pwr.edu.pl (F.F.)
[2] Faculty of Science and Engineering, Kingston University, London SW15 3DW, UK; augousti@kingston.ac.uk
* Correspondence: Andrzej.Radosz@pwr.edu.pl; Tel : +48-660-112-122

Received: 30 July 2018; Accepted: 22 August 2018; Published: 27 August 2018

Abstract: A black hole in a Schwarzschild spacetime is considered. A transformation is proposed that describes the relationship between the coordinate systems exterior and interior to an event horizon. The application of this transformation permits considerations of the (a)symmetry of a range of phenomena taking place on both sides of the event horizon. The paper investigates two distinct problems of a uniformly accelerated particle. In one of these, although the equations of motion are the same in the regions on both sides, the solutions turn out to be very different. This manifests the differences of the properties of these two ranges.

Keywords: black hole; event horizon; asymmetry

---

1. Introduction

Outside the horizon of a black hole the Killing vector is time-like and it becomes space-like inside the horizon—this means that energy conservation is broken (see e.g., [1]). Such a conversion of the Killing vector results in a lot of interesting phenomena, with the Hawking radiation and the information paradox being the most prominent among them [2,3]. The status of these two outcomes arising from the presence of the event horizon is, in a sense, similar: both appeared mainly due to the work of Hawking and both of them are still debated [4]. More recent studies have considered a range of interesting questions. One of the latest was the supposition that black holes (BH) do indeed have hair [5]; another interesting outcome was the claim that the volume of a BH may be infinite [6]. In view of the argument that there is not enough space inside a black hole to store arbitrarily large amounts of information, this discovery turns out to be a contribution that is strictly associated with the problem of information stored and lost within an event horizon [7]. Following a proposal by Unruh [8] of development of gravity analogues, condensed, soft matter or photonic black hole-like systems that develop an event horizon have been experimentally realized [9]. A recent experiment has led to a Kerr-like BH, including an ergosphere, not just an event horizon, opening up the possibility of experimental observations of the Penrose effect [10].

The interior of a black hole itself has been the subject of fewer but more diverse considerations. Hamilton et al. [11] presented a discussion on the question of (spectroscopic) vision inside the horizon of a Schwarzschild BH. In another approach Hamilton and Leslie [12] considered the exterior and interior of a BH within a so-called river model that was based on the use of a Painleve-Gullstrand co-moving frame. It is well-known that the static spacetime outside a BH's horizon turns out to be dynamic inside the outer horizon in the situation when two horizons are formed. Such a dynamic interior of a BH may be viewed as a cosmological model [13].

The aim of this paper is to study the properties of the interior of a Schwarzschild BH, i.e., which is isotropic and static outside the horizon. The investigations are presented from a special perspective. One finds a formal, specific symmetry between the exterior and interior ranges, even though the properties of these two regions are found to be completely different. We illustrate the differences using two problems that are related to the behavior of a test particle uniformly accelerated outside and inside the horizon. In the first example, the particle moves radially outside the horizon; inside the horizon, corresponding motion takes place along a homogeneity axis (see below). It is found then that although the equations of motion are the same in both regions the solutions differ significantly. In the other example, a test particle follows a circular trajectory belonging to the photon sphere [14] and inside the horizon the test particle moves along a photon sphere analogue.

The paper is organized as follows. In Section 2, we present a transformation between coordinate systems outside and inside the horizon of the Schwarzschild black hole. Section 3 is devoted to a thorough investigation of the problem of uniformly accelerated motion along a straight line: radially outside the horizon, and along the homogeneity direction inside the horizon. In Section 4, we consider the case of uniform acceleration for motion along a circle on a photon sphere outside horizon and its analogue inside the horizon. A discussion and final remarks are provided in the last section.

2. Systems of Co-Ordinates in Space-Time with a Horizon

Arbitrary co-ordinate system constitute a diffeomorphism Φ from an open set U of the space-time M into an open set U' of the real Euclidean flat space R^n, where n is the dimension of M:

$$\Phi : U \subset M \to U' \subset R^n$$

$$\Phi(p) = (x^1, \ldots x^n) \text{ where } p \in U. \tag{1}$$

We will consider a spherically symmetric spacetime with a black hole defined by a horizon. The horizon H, which is a global property of such a spacetime, divides M into an exterior M_+ and an interior M_-:

$$M = M_- \cup H \cup M_+. \tag{2}$$

On M_+ (which is static) one can introduce a coordinate system which in spherical coordinates (r, θ, φ) on R^3 takes the form:

$$\Phi_+ : M_+ \to R^4$$

$$\Phi_+(p) = (t, r, \theta, \varphi) = (x^\mu), \tag{3}$$

where $r > r_g$ and r_g is a (gravitational) radius of the horizon H. The metric ds_+^2 on M_+ in this framework has the form

$$ds_+^2 = \left(1 - \frac{r_g}{r}\right)dt^2 - \left(1 - \frac{r_g}{r}\right)^{-1}dr^2 - r^2 d\Omega^2. \tag{4}$$

This is the well-known Schwarzschild space-time with $r_g = 2M$, and $d\Omega^2 = d\theta^2 + \sin^2\theta \, d\varphi^2$ is a unit metric on S^2.

On M_- one can introduce the following coordinate system:

$$\Phi_- : M_- \to R^4$$

$$\Phi_-(p) = (T, R, \theta, \varphi) = (X^\alpha), \tag{5}$$

where $0 < T < r_g$. The metric ds_-^2 on M_- in this coordinate system has the following form (see [11])

$$ds_-^2 = \frac{dT^2}{-1 + \frac{r_g}{T}} - \left(\frac{r_g}{T} - 1\right)dR^2 - T^2 d\Omega^2. \tag{6}$$

Hence, on M, there are two coordinate systems with two metrics $ds_+^2 = g_{\alpha\beta}(x)dx^\alpha dx^\beta$ and $ds_-^2 = h_{\alpha\beta}(X)dX^\alpha dX^\beta$. The transformation $\mathbb{T}$ between these two systems of coordinates has the form:

$$\mathbb{T}: M_+ \to M_-, \tag{7}$$

where $\mathbb{T}$ is given by the matrix:

$$\mathbb{T} = \begin{pmatrix} 0 & 1 & 0 & 0 \\ 1 & 0 & C & 0 \\ 0 & 0 & 1 & 0 \\ 0 & 0 & 0 & 1 \end{pmatrix} = \mathbb{T}^{-1}. \tag{8}$$

It acts as follows:

$$\mathbb{T}x = X,$$

$$T = r, \quad R = t, \quad \theta = \theta, \quad \varphi = \varphi, \tag{9}$$

and

$$h(X) = \mathbb{T}g(\mathbb{T}x)\mathbb{T}, \tag{10}$$

where $h(X) = (h_{\alpha\beta}(X))$ and $g(\mathbb{T}x) = (g_{\alpha\beta}(\mathbb{T}x))$.

There are two Killing vectors manifesting symmetry properties in the exterior M_+ and interior M_- of the horizon. In M_+, there is a time-like one, ∂_t representing energy conservation due to the static character of the spacetime and a space-like one, ∂_φ reflecting angular momentum conservation. In M_-, there is a space-like one, ∂_R reflecting R-momentum conservation due to homogeneity along the R-axis and another space-like one, ∂_φ representing angular momentum conservation.

The exterior of the Schwarzschild black hole, M_+ is a static spacetime and its interior, M_- is a dynamic spacetime. One can introduce a class of static observers in the former and a class of resting observers in the latter case. The velocity vector of such observers has only a temporal non-vanishing coordinate $U = \alpha \partial_0$. It appears natural to label them as:

(a) t-observers

$$U_t = \frac{1}{\sqrt{g_{tt}}} \partial_t \tag{11}$$

outside the horizon and

(b) T-observers

$$U_T = -\frac{1}{\sqrt{h_{TT}}} \partial_T \tag{12}$$

inside the horizon.

3. Uniformly Accelerated Motion along Straight Line

In this section we will consider the problem of a uniformly accelerated test particle in the exterior and interior of the horizon. The motion of such a test particle will be confined to the x^0, x^1 and X^0, X^1 hyperplanes, respectively. So, one can discuss the case of a $1 + 1$ dimensional spacetime

$$ds^2 = f(d\zeta^0)^2 - f^{-1}(d\zeta^1)^2 \tag{13}$$

such that

(a)   Outside the horizon, $\zeta^\alpha \equiv x^\alpha$ and $f = g_{tt}$ is a function of spatial coordinate ζ^1
(b)   Inside the horizon, $\zeta^\alpha \equiv X^\alpha$ and $f = h_{TT}$ is a function of temporal coordinate ζ^0.

The components of the velocity vector u

$$u = u^\alpha \partial_\alpha \tag{14}$$

of the test particle, u^α will depend on ξ^1 in the case (a) and on ξ^0 in case (b).

An acceleration vector field a for u is:

$$a = \nabla_u u \tag{15}$$

and one obtains the following equations for its components:

(a)   outside the horizon

$$a^t = -g_{rr} u^r \frac{d}{dr}(g_{tt} u^t) \tag{16}$$

$$a^r = g_{tt} u^t \frac{d}{dr}(g_{tt} u^t) \tag{17}$$

and

(b)   inside the horizon

$$a^T = h_{RR} u^R \frac{d}{dT}(h_{RR} u^R) \tag{18}$$

$$a^R = -h_{TT} u^T \frac{d}{dT}(h_{RR} u^R) \tag{19}$$

Hence, one finds that these two pairs of equations are transformed into each other under the interchange of the temporal and spatial coordinates, $t \leftrightarrow R$ and $r \leftrightarrow T$ as emphasized in the former section (see Equation (9)). Uniform acceleration is defined by the condition:

$$a^2 = f(a^0)^2 - f^{-1}(a^1)^2 = -\alpha^2 \tag{20}$$

where $\alpha = const$. The equations for the world line of the test particle are then derived, as follows.

(a)   Outside the horizon

$$\frac{d}{dr}(g_{tt} u^t) = \pm\alpha. \tag{21}$$

Thus, the world line $\gamma = \{t(\tau), r(\tau)\}$ of the uniformly accelerated particle is given in the case of a static metric by the integral curve of the vector field:

$$u^t = \frac{dt}{d\tau} = \frac{1}{g_{tt}}\left(E \pm \alpha \int dr\right) \tag{22}$$

$$u^r = \frac{dr}{d\tau} = \pm\sqrt{\left(E \pm \alpha \int dr\right)^2 - g_{tt}} \tag{23}$$

where E is an integration constant.

(b)   Inside the horizon

$$\frac{d}{dT}(h_{RR} u^R) = \pm\alpha. \tag{24}$$

In this case of a dynamic spacetime the world line of the test particle is given, as follows:

$$u^R = \pm\frac{1}{h_{RR}}\left(E' \pm \alpha \int dT\right) \tag{25}$$

$$u^T = \sqrt{\left(E' \pm \alpha \int^r dT\right)^2 - h_{RR}}.$$ (26)

Applying the fact that inside the horizon

$$g_{tt} = 1 - \frac{r_g}{r} = h_{RR}(r)$$ (27)

one finds that the world line equations for the outer and inner horizon regions, Equations (22) and (23) and Equations (25) and (26) take the same form:

$$\dot{t} = \frac{E + \alpha r}{1 - \frac{r_g}{r}}$$ (28)

$$\dot{r} = \sqrt{(E + \alpha r)^2 - \left(1 - \frac{r_g}{r}\right)}.$$ (29)

Here, we will examine the following point: how does the speed of a uniformly accelerated test particle change with respect to the static observers? In general a relative velocity four-vector w of u', as measured by another observer characterized by velocity u at the same event $p = (x^\alpha)$ is given by (see [15])

$$w = \frac{u'}{(u, u')} - u,$$ (30)

where (u, u') is the scalar product of the two velocity four-vectors, u and u', $u^2 = u'^2 = 1$. Velocity w is orthogonal to u, $(w, u) = 0$, i.e., w is a space-like vector,

$$w^2 = \frac{1}{(u, u')^2} - 1 < 0.$$ (31)

Then, the squared speed v^2 is given as:

$$v^2 = -w^2 > 0.$$ (32)

3.1. Black Hole Exterior

In the exterior of the Schwarzschild black hole the spacetime is static and static observers denoted as t-o are characterized by their velocity vector U_t (see Equation (11)). Such an observer measures the speed v of a nearby passing uniformly accelerated (ua) test particle,

$$u_{ua} = \dot{t}\partial_t + \dot{r}\partial_r.$$ (33)

Hence, for $u = U_t$, and $u' = u_{ua}$, one applies the procedure described above to find:

$$w^2 = -\frac{\dot{r}^2}{\left(g_{tt}\dot{t}\right)^2}.$$ (34)

Using Equations (28) and (29), we obtain

$$v^2 = \frac{(E + \alpha r)^2 - g_{tt}}{(E + \alpha r)^2}.$$ (35)

This case of a uniformly accelerated test particle initially at rest, $r = 2r_g$ is illustrated in Figure 1.

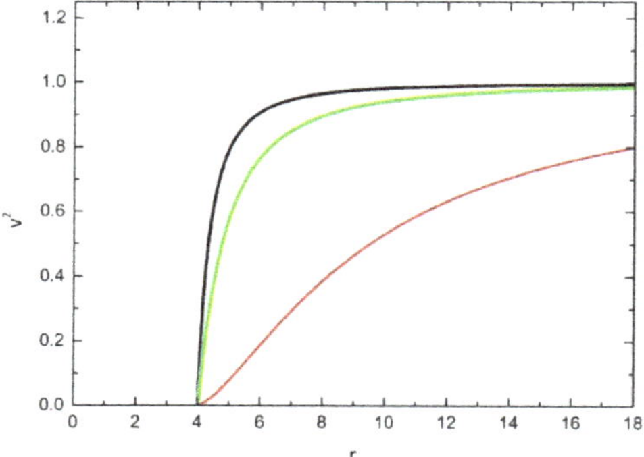

Figure 1. Squared speed v^2 (Equation (35)) of a uniformly accelerated test particle initially at rest at $r_0 = 2r_g \equiv 4$ (in this case $r_g \equiv 2$) escaping radially from the gravitational field for different values of α = 0.1 (red), $\alpha = 0.5$ (green), $\alpha = 1$ (black).

3.2. Black Hole Interior

Inside the horizon, $h_{TT} = \left(\frac{r_g}{T} - 1\right)^{-1}$. A class of resting (or co-moving, see below) observers is determined by (12)

$$U_T = -\sqrt{\frac{r_g}{T} - 1}\,\partial_T. \tag{36}$$

Using Equations (26), (32), and (36), one obtains the squared speed $\tilde{v}^2$ of the uniformly accelerated test particle inside the horizon:

$$\tilde{v}^2 = \frac{(E' + \alpha T)^2}{(E' + \alpha T)^2 - \left(1 - \frac{r_g}{T}\right)}. \tag{37}$$

Using the relation $T = r$ inside the horizon, one can express the squared speed $\tilde{v}^2$ as a function of a temporal coordinate, r. In Figure 2, the squared speed of the uniformly accelerated test particle initially at rest is illustrated by applying expression (37). The speed initially increases, reaches some (non-universal) maximal value and finally decreases to zero at the ultimate singularity, $r \to 0$.

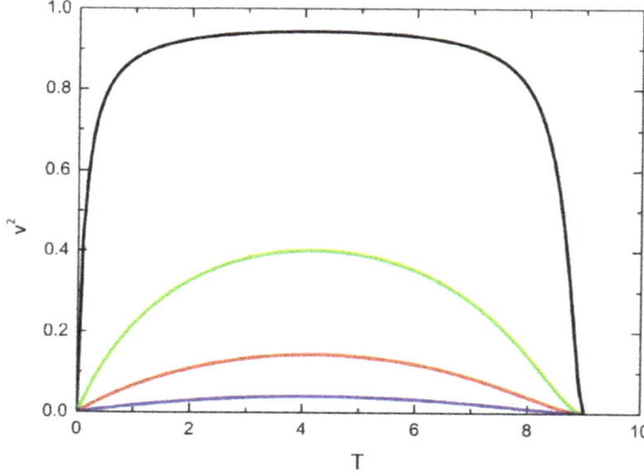

Figure 2. Squared speed $\tilde{v}^2$ (Equation (37)) of a test particle initially, $T_0 = 0.9\, r_g$ at rest (in this case $r_g \equiv 10$) uniformly accelerated along homogeneity axis $R(=t)$ for different values of $\alpha = 0.05$ (blue), $\alpha = 0.1$ (red), $\alpha = 0.2$ (green), $\alpha = 1$ (black).

4. Uniform Acceleration on a Photon Sphere and on Its Analogue inside the Horizon

A photon sphere outside the horizon, $r > r_g$, has a radius, $r_{psh} = \frac{3}{2}r_g$. It consists of null geodesic circles, $k = k^t \partial_t + k^\varphi \partial_\varphi$, $k^2 = 0$ $(r = r_{psh})$. Its analogue inside the horizon, $r < r_g$, consists of planar null geodesics having two non-vanishing components, $k = k^r \partial_r + k^\varphi \partial_\varphi$. However, in this case, the radius, r of such a "sphere" diminishes from r_g to C. We shall consider the problem of a uniformly accelerated test particle on a circle belonging to the photon sphere and on a "circle" belonging to the photon sphere analogue.

4.1. Black Hole Exterior

In a case of a test particle moving outside the horizon along a circle of radius $r_0 \geq r_{psh}$ the velocity vector is two-component

$$u = u^t \partial_t + u^\varphi \partial_\varphi, \tag{38}$$

where without loss of generality, we have considered a circle in an equatorial plane, $\theta = \pi/2$. An acceleration vector field $a = \nabla_u u$ in the case of motion with constant speed is found to be $a = a^r \partial_r$ (see [14]).

$$a^r = -\frac{1}{2}\left[(u^t)^2 \frac{dg_{tt}}{dr} + (u^\varphi)^2 \frac{dg_{\varphi\varphi}}{dr} \right]. \tag{39}$$

Applying a normalization condition, $u^2 = 1$, one finds that the acceleration of a photon sphere, $r_0 = r_{psh} = \frac{3}{2}r_g$, turns out to be speed independent [14]

$$a^r = -\frac{1}{2}\frac{1}{g_{tt}}\frac{dg_{tt}}{dr} = \frac{2}{3r_g}. \tag{40}$$

Hence, in this case, one can consider a generalization of the twin-paradox and other features, such as specific non-geodesic motion [16].

4.2. Black Hole Interior

Motion along a photon sphere analogue in this case, $r < r_g$ is as follows.
The two-component velocity vector of a test particle takes the form:

$$u = u^r \partial_r + u^\varphi \partial_\varphi, \tag{41}$$

and its components will be r-dependent. An acceleration vector field $a = \nabla_u u$ is found to be two-component,

$$a = a^r \partial_r + a^\varphi \partial_\varphi, \tag{42}$$

and its components are:

$$a^r = -\frac{u^\varphi}{g_{rr}}\frac{d}{dr}(g_{\varphi\varphi}u^\varphi), \tag{43}$$

$$a^\varphi = \frac{u^r}{g_{\varphi\varphi}}\frac{d}{dr}(g_{\varphi\varphi}u^\varphi). \tag{44}$$

Uniform acceleration is defined by the condition (see Equation (13))

$$a^2 = g_{rr}(a^r)^2 + g_{\varphi\varphi}(a^\varphi)^2 = -\beta^2 = const. \tag{45}$$

The equations for the world line of the test particle are then derived as follows. Applying Equations (43)–(45), one finds,

$$\frac{d}{dr}(g_{\varphi\varphi}u^\varphi) = \pm\beta\sqrt{-g_{rr}g_{\varphi\varphi}}. \tag{46}$$

Thus, the world line $\gamma = \{r(\tau), \varphi(\tau)\}$ of the test particle uniformly accelerated on a sphere of ever diminishing radius, is given by an integral curve of the vector field u:

$$u^r = \frac{dr}{d\tau} = \pm \frac{1}{\sqrt{g_{rr}}} \left(1 - \frac{1}{g_{\varphi\varphi}} \left(C \pm \beta \int \sqrt{-g_{rr}g_{\varphi\varphi}} dr \right)^2 \right), \tag{47}$$

$$u^\varphi = \frac{d\varphi}{d\tau} = \frac{1}{g_{\varphi\varphi}} \left(C \pm \beta \int \sqrt{-g_{rr}g_{\varphi\varphi}} dr \right), \tag{48}$$

where C is an integration constant. Here, we are interested in the following question (see above): how does the speed of a uniformly accelerated test particle change with respect to the static (resting) observers? Similarly, as in the former section, the speed is measured by (resting) T-observers, Equation (36). In this case, one finds:

$$\tilde{v}^2 = 1 - \frac{1}{g_{rr}(u^r)^2} = \frac{\left(C \pm \beta \int \sqrt{-g_{rr}g_{\varphi\varphi}} dr \right)^2}{\left(C \pm \beta \int \sqrt{-g_{rr}g_{\varphi\varphi}} dr \right)^2 - g_{\varphi\varphi}}. \tag{49}$$

The meaning of the result (49) is: the speed of the uniformly accelerated test particle, of total acceleration β tends to the speed of light as the ultimate singularity, $r \to 0$ approaches, as $g_{\varphi\varphi} = -r^2 \to 0$.

5. Discussion and Final Remarks

In order to investigate the properties of the interior of a Schwarzschild black hole one can apply a specific symmetry to the coordinate systems of its exterior and interior. The meaning of such a symmetry is that the radial and temporal coordinates interchange their roles. This symmetry may be regarded as a justification of the fact that one can apply a Schwarzschild system of coordinates, (t, r, θ, φ) both outside as well as inside the horizon. It is an almost trivial notion, but it is important to remember the existence of a singularity in this system, which is the horizon itself.

In the case of radial motion outside the horizon and motion along the direction of homogeneity inside the horizon, one finds that the equations of motion on both sides of the horizon obey the symmetry: they are transformed into each other under transformation $\mathbb{T}$ (7–10). This is because the motion takes place in the hyper-planes $t - r$ and $T - R$ that are transformed into each other under $\mathbb{T}$. However, the properties of the solutions are very different. Outside the horizon, a radially accelerated particle departs with ever increasing speed (if the acceleration is larger than some critical value). The speed of a test particle uniformly accelerated along the homogeneity axis inside the horizon, whose equation of motion is the same as the one outside the horizon, initially increases but then decreases to zero when approaching the ultimate singularity. This means that transformation $\mathbb{T}$ does not lead to particular physical consequences (at least in this case).

In the case of circular motion on a photon sphere and "circular" motion on a photon sphere analogue, outside and inside the horizon, respectively, the equations of motion are (rather obviously) different. Apart from the fact of the existence of a photon sphere and its analogue inside the horizon, there are no features that are in common for accelerated motion of a test particle in such orbits. The radial acceleration during circular motion of the test particle on the photon sphere is independent of its speed (which is a well-known result); the speed of a test particle following accelerated motion along a circle belonging to the photon sphere analogue (of ever decreasing radius) increases to the speed of light when approaching the ultimate singularity.

Having in mind an (intimate) symmetry $\mathbb{T}$ between the exterior and interior of the Schwarzschild black hole, leading in the simplest case to the same equation of motion, one may wonder how it arises that the properties of the solutions are so different. The answer is: it is because r and t, spatial (radial) and temporal coordinates, respectively, outside the horizon interchange their roles, becoming temporal and spatial (in the direction of homogeneity) coordinates inside the horizon. Such an exchange of roles has a deeper consequence: outside the horizon, spacetime is spherically symmetric and static and inside the horizon it is no longer static, it is dynamically changing (i.e., an "anisotropic cosmology", see [13]), but homogeneous along one of its spatial directions.

It should be noted that our discussion has been presented within a specific system of coordinates, namely Schwarzschild coordinates outside the horizon, revealing a singularity, namely the horizon itself. One may ask then: what is the significance of these results? The answer is as follows.

When considering the case of uniformly accelerated motion along a straight line (Section 3), one finds acceleration vectors outside, Equations (16) and (17) and inside, Equations (18) and (19) the horizon. These quantities (being vectors) may be transformed in a covariant way to another (arbitrary) coordinate system, e.g., a Kruskal-Szekeres coordinate system (see e.g., [1]), one free of singularities. Transformation between Schwarzschild, t, r and Kruskal-Szekeres, u, v coordinates,

$$\frac{\partial(u,v)}{\partial(t,r)} = \begin{pmatrix} \frac{1}{4M}v & \frac{1}{4M}\frac{1}{1-\frac{2M}{r}}u \\ \frac{1}{4M}u & \frac{1}{4M}\frac{1}{1-\frac{2M}{r}}v \end{pmatrix}$$

enables us to confirm that symmetry of the equations of motion Equations (16)–(19) found in the former system is also preserved in the latter one.

The condition of a uniform acceleration, (20) is in fact independent of coordinate system (as it refers to a scalar product, i.e., a squared acceleration vector). The world line itself is characterized by velocity vectors, outside, Equations (22) and (23) and inside, Equations (25) and (26) the horizon; this characteristic, velocity vector, is also transformed in a covariant way to other systems of coordinates.

A physical quantity illustrating the properties of the interior of a Schwarzschild black hole, such as the (squared) speed of a uniformly accelerated test particle, as measured by a resting T-observer inside the horizon is expressed in terms of a scalar product, Equations (30)–(32) and (37); hence, it is independent of the coordinate system.

The case of a test particle following a circular orbit and its analogue inside the horizon may be analyzed in a similar manner leading to a similar outcome. The conclusion that the squared speed inside the horizon of an accelerated test particle tends to the value of the speed of light at the ultimate singularity is obviously a coordinate independent result.

One should underline however that the choice of the specific system of coordinates allows for one to (easily) solve Equation (21), leading to the explicit form Equations (28) and (29); solving the equation of motion in another coordinate system may not be such an easy task.

Author Contributions: P.G.: The idea of (a)symmetry, uniform acceleration, calculations; A.A.: Conceptual discussions, manuscript preparation and editing; F.F.: Numerical support, preparations of illustrations; A.R.: Discussing and developing the concepts of uniformly accelerated test particles as viewed by resting observers, calculations.

Funding: Department of Quantum Technologies, Wroclaw University of Science and Technology.

Conflicts of Interest: The authors declare no conflict of interest.

References

1. Hartle, J.B. *Gravity*; Addison Wesley: Boston, MA, USA, 2003; Chapter 12, p. 13.
2. Hawking, S.W. Black hole explosions? *Nature* **1974**, *248*, 30–31. [CrossRef]
3. Hawking, S.W. Breakdown of predictability in gravitational collapse. *Phys. Rev. D* **1976**, *14*, 2460–2473. [CrossRef]
4. Almheiri, A.; Marolf, D.; Polchinski, J.; Sully. J. Black holes: Complementarity or firewalls? *J. High Energy Phys.* **2013**, *62*, 1–19. [CrossRef]
5. Hawking, S.W.; Perry, M.; Strominger, A. Soft Hair on Black Holes. *Phys. Rev. Lett.* **2016**, *116*. [CrossRef] [PubMed]
6. Christodoulou, M.; Rovelli, C. How big is a black hole? *Phys. Rev. D* **2015**, *91*, 1–7. [CrossRef]

7.	Mathur, S. A model with no firewall. *arXiv* **2015**, arXiv:1506.04342.
8.	Unruh, W.G. Experimental Black-Hole Evaporation? *Phys. Rev. Lett.* **1981**, *46*, 1351–1353. [CrossRef]
9.	Nguyen, H.S.; Gerace, D.; Carusotto, I.; Sanvitto, D.; Galopin, E.; Lemaître, A.; Sagnes, I.; Bloch, J.; Amo, A. Acoustic Black Hole in a Stationary Hydrodynamic Flow of Microcavity Polaritons. *Phys. Rev. Lett.* **2015**, *114*. [CrossRef] [PubMed]
10.	Vocke, D.; Maitland, C.; Prain, A.; Biancalana, F.; Marino, F.; Faccio, D. Rotating black hole geometries in a two-dimensional photon superfluid. *Optica* **2018**, in press.
11.	Hamilton, A.J.S.; Polhemus, G. Stereoscopic visualization in curved spacetime: Seeing deep inside a black hole. *New J. Phys.* **2010**, *12*, 123027–123052. [CrossRef]
12.	Hamilton, A.J.S.; Lisle, J.P. The river model of black holes. *Am. J. Phys.* **2008**, *76*, 519–532. [CrossRef]
13.	Doran, R.; Lobo, F.S.N.; Crawford, P. Interior of a Schwarzschild black hole revisited. *Found. Phys.* **2008**, *38*, 160–187. [CrossRef]
14.	Abramowicz, M.A.; Bajtlik, S.; Kluzniak, W. The twin paradox on the photon sphere. *Phys. Rev. A* **2007**, *75*. [CrossRef]
15.	Bolos, V.J. Intrinsic Definitions of "Relative Velocity" in General Relativity. *Comm. Math. Phys.* **2007**, *273*, 217–236. [CrossRef]
16.	Abramowicz, M.A.; Bajtlik, S. Adding to the paradox: The accelerated twin is older. *arXiv* **2009**, arXiv:0905.2428v1 2009.

MDPI

St. Alban-Anlage 66

4052 Basel

Switzerland

Tel. +41 61 683 77 34

Fax +41 61 302 89 18

www.mdpi.com

Symmetry Editorial Office

E-mail: symmetry@mdpi.com

www.mdpi.com/journal/symmetry